Best of Silver's Integration Bee Problems (from 2019-2023)

Yuepeng Alex Yang

Through all the Years of Silver's Problem Writing

I am very thankful to a lot of people. To my family for letting me pursue my dreams, my best friend for getting me into speed integration, and to the organizers of MUSA of UC Berkeley, UCONN, Berkeley Math Tournament, Stanford Math Tournament, and The Archimedeans (UK University Integration Bee) for accepting me as a problem writer for Integration Bees. This book would never exist without them.

Contents

Preface

Wasabi!!! This is a problem book for those who wants to learn sneaky fast integration techniques as well as using this book as a reference to help readers get inspired to create their own integration bee problems. The following book contains a list of all archived integration bee problems from 2019-2023. I have been cherry picking what I think is super interesting and some of the best integration bee problems made for all levels: beginners, intermediates, advanced, and the olympiads.

The book will first list all of the integration bee problems so that the readers can try them out on their own. The problems are cut into three sets. The first set are integration bee problems made during my high school years. The second set are integration bee problems made for university competitions and math tournaments. The third set are integration bee problems made for training purposes and mock exams. After the list of three sets of problems, the solutions are provided respectively.

If you're currently training for an integration bee, I'd suggest you to go watch my Youtube series: Integration Bee Training. There are three playlists you can easily find: for beginners, for intermediates, and for advanced. This book is not beginner friendly and is leaned towards those who are experienced speed integrators or intermediate level integration bee competitors. This book is mainly just to showcase and provide unique challenging integrals for those audiences.

Enjoy the integrals!!!

Chapter 1

Problems from Silver's Evil High School Integrals

1.
$$\int \frac{d}{dx}\left[\frac{x^2e^x}{\ln(x)}\right]\cdot\frac{\ln(x)}{x^2e^x}dx$$

2.
$$\int x^{\frac{\ln(\ln(x))}{\ln(x)}}dx$$

3.
$$\int \frac{e^{x+1+e^{x+1}}}{1+e^{1+e^{1+x}}}dx$$

4.
$$\int \sqrt[4]{\cos^4(x)-\cos(2x)}dx$$

5.
$$\int \left(\frac{\cos(x)}{\sin(x)+\cos(x)}\right)^2 dx$$

6.
$$\int \tan(x)\tan(2x)dx$$

7.
$$\int \sqrt{1-2\sin(2x)+3\cos^2(x)}dx$$

8.
$$\int \frac{\ln(x)-1}{x^2-\ln^2(x)}dx$$

9.
$$\int \sqrt{\frac{x}{\ln(x)}}\left(1-\frac{1}{3\ln(x)}\right)dx$$

10.

$$\int \frac{\cos^2(x)+1}{\cos(x)\sqrt{\cos(x)}}dx$$

11.

$$\int \frac{dx}{\sqrt{x}\sqrt{1+\sqrt{x}}\sqrt{1+\sqrt{1+\sqrt{x}}}}$$

12.

$$\int (x+3)\sqrt{xe^x}dx$$

13.

$$\int \left(\frac{(x+1)^{2019}(x^2+x+1)^{2019}}{(x^3+1)^{2019}}\right)^{2019}\left(\frac{(x-1)^{2019}(x^2-x+1)^{2019}}{(x^3-1)^{2019}}\right)^{2019} dx$$

14.

$$\int \left(\left(e^{e^{e^x}}+1\right)e^{e^x}+1\right)e^x dx$$

15.

$$\int \frac{\sec(x)+x+\tan(x)}{(\sec(x)+\tan(x))^2}dx$$

16.

$$\int \frac{x}{x-1+e^{1-x}}dx$$

17.

$$\int \frac{\ln(1+\sqrt{1-x^2})-\ln(x)}{x\sqrt{1-x^2}}dx$$

18.

$$\int \frac{\sin(\sin(x)-x)}{\cos(\sin(x))}dx$$

19.

$$\int \sqrt{x(x+1)(x+2)(x+3)+1}\, dx$$

20.

$$\int e^{e^x\cos(x)+x}(\sin(x)-\cos(x))dx$$

21.

$$\int \sqrt{1+\sec^2(x)+2\sqrt{2}\tan(x)}\, dx$$

22.

$$\int_0^1 \frac{\sqrt{1+x}}{\sqrt{1+\sqrt{1-x^2}}}dx$$

23.

$$\int \frac{dx}{e^{-\sqrt{x}}+\sqrt{x}-1}$$

24.

$$\int_0^{\frac{\pi}{3}} \left(\frac{\sec(x)+\tan(x)}{\csc(x)+\cot(x)}\right)(\sec(x)+\csc(x))dx$$

25.

$$\int (x-2)e^{\left(\frac{1-x-xe^x}{e^x}\right)}dx$$

26.

$$\int \frac{x-1}{\sqrt{xe^x}}dx$$

27.

$$\int e^{\sin(x)}e^{\cos(x)}\cos(2x)dx$$

28.

$$\int \frac{\sec(x)\tan(x)-\sec^2(x)}{\sqrt{\sec(x)+\tan(x)}}dx$$

29.

$$\int_1^2 e^{\sec^{-1}(x)}\left(\frac{\sqrt{x^2-1}+1}{x}\right)dx$$

30.

$$\int \frac{\tan(2x)(\sin^{14}(x)+\cos^{14}(x))}{(\sin^4(x)+\cos^4(x))(\sin^8(x)+\cos^8(x))}dx$$

31.

$$\int \frac{e^{e^x}+1}{e^{e^x-x}+1}dx$$

32.

$$\int \left(\frac{\cos(x)}{\cos(x+\sin(x)\cos(x))}\right)^2 dx$$

33.

$$\int \frac{\cos(x)}{\sqrt{2}\cos(\frac{x}{2})+1}dx$$

34.

$$\int \frac{1-x}{\sqrt{e^{2x}-x^2}}dx$$

35.

$$\int \frac{1}{x^2}e^{\sec^{-1}(x)}dx$$

36.

$$\int \frac{1+\frac{1+\frac{1\frac{1}{2\sqrt{x}}}{2\sqrt{x+\sqrt{x}}}}{2\sqrt{x+\sqrt{x+\sqrt{x}}}}}{2\sqrt{x+\sqrt{x+\sqrt{x+\sqrt{x}}}}}dx$$

37.

$$\int_2^6 \sqrt{x - \frac{1}{x - \frac{1}{x - \dots}}} dx$$

38.

$$\int x^x \left(\ln^2(x) + \ln(x) + \frac{1}{x} \right) dx$$

39.

$$\int \frac{x^x}{x^x + \sqrt{x}} + \frac{x^{1-x}}{x^{1-x} + \sqrt{x}} dx$$

40.

$$\int x^e(e^x + e^{-x}) + x^{e-1}(e^{x+1} - e^{1-x}) dx$$

41.

$$\int \sqrt{1 + x\sqrt{1 + (x+1)\sqrt{1 + (x+2)(x+4)}}}\, dx$$

42.

$$\int \left(\frac{\cos(x)}{\sqrt{\sin(x)}} - \frac{\sin(x)}{\sqrt{\cos(x)}} \right) \left(\frac{\sin(x) - \cos(x)}{\sqrt{\sin(x)} - \sqrt{\cos(x)}} \right) dx$$

43.

$$\int \frac{(x+1)\sin(x) + (x-1)\cos(x)}{x(\sin(x) - \cos(x))} dx$$

44.

$$\int \tan^{-1}(\sec(x) + \tan(x)) dx$$

45.

$$\int \frac{\sin^2(x)}{2x - \sin(2x)} dx$$

46.

$$\int \sec^2(x)\tan(x)\ln(\sec(x)) dx$$

47.

$$\int \frac{x+1}{x\sqrt{xe^x - 1}} dx$$

48.

$$\int \frac{x}{x\tan(x) + 1} + \frac{1}{\sqrt{x}\tan(\sqrt{x}) + 1} dx$$

49.

$$\int \sqrt{x}\ln(x)\sqrt{\ln(x) - 1}\, dx$$

50.

$$\int \frac{\tan(x)}{\sqrt{\sec^4(x) - 1}} dx$$

51.

$$\int \frac{\ln(e^x - 1) - x}{e^x - 1} dx$$

52.

$$\int \frac{\ln((x+1)^{(x+1)} x^x)}{x(x+1)} dx$$

53.

$$\int \frac{e^{\sec(x) - \tan(x)}}{1 + \sin(x)} dx$$

54.

$$\int \left(1 - \frac{1}{x^2}\right) e^{\sqrt{x} + \frac{1}{\sqrt{x}}} dx$$

55.

$$\int_{-\frac{\pi}{4}}^{\frac{\pi}{4}} \frac{\sqrt{\sec(2x)}}{\cos(x) + \sqrt{\cos(2x)}} dx$$

56.

$$\int_0^1 \frac{x\sqrt{\sqrt{1-x} + \sqrt{1+x}}}{\sqrt{1-x^2}} dx$$

57.

$$\int (256^x + 2^{x+1})^7 dx$$

58.

$$\int_0^1 \frac{2^x}{(\sqrt{5}+1)^x + (\sqrt{5}+3)^x} dx$$

59.

$$\int_0^{\frac{\pi}{3}} \frac{1 + x\tan(x)}{\sqrt{x\cos(x)}} dx$$

60.

$$\int \frac{x^2}{(x^2 - 2x + 4)(x^2 + 3x + 9)(x^2 - x - 6)} dx$$

61.

$$\int \frac{\sqrt{\cos(2x)}}{\sin(x)} dx$$

62.

$$\int \frac{2x^3 - 1}{x^2 + (x^3 + 1)^2} dx$$

Chapter 2

Problems made for University Competitions

1.

$$\int \cos(\sin(x) - x) + \cos(\sin(x) + x)dx$$

2.

$$\int x^{\frac{x^{\frac{x}{\ln(x)}}}{\ln(x)} + \frac{x}{\ln(x)}} dx$$

3.

$$\int_{-\infty}^{\infty} \frac{\sin(2x)\cos(\frac{1}{x}) - \sin(\frac{1}{x})\cos(2x)}{2x^3 - x} dx$$

4.

$$\int \left(e^{x^3 - x}\right)\left(3x^3 - x + 1\right) dx$$

5.

$$\int_{-\infty}^{\infty} \frac{e^{x+\frac{1}{x}}}{e^{\frac{2}{x}} + e^{2x}} dx$$

6.

$$\int \frac{dx}{\cos x\sqrt{\sin x}}$$

7.

$$\int_{1}^{\sqrt{2}} \frac{2x(\sqrt{x^2-1} + \sec^{-1}(x))}{(x^2+1)^2} dx$$

8.

$$\int_{2}^{3} \frac{dx}{\sqrt{1 - \frac{2}{x}}}$$

9.

$$\int_{0}^{\frac{\pi}{4}} \ln\left(\sec x + \tan x\right)\left(\cos^2 x - \frac{1}{2}\right) dx$$

10.
$$\int \frac{dx}{\sqrt{\sin(x)\left(\cos(x)+1\right)}}$$

11.
$$\int \frac{dx}{\sin x\left(\ln\left(\sin x\right)-\ln\left(1+\cos x\right)\right)}$$

12.
$$\int_2^e \ln\left(\sqrt[x]{e\sqrt[x]{e\sqrt[x]{e\sqrt[x]{e\sqrt[x]{\ldots}}}}}\right)dx$$

13.
$$\int \frac{\tan(x)}{\sqrt{\cos(2x)}}dx$$

14.
$$\int_0^1 (x^{2022}+x)^{2020}dx$$

15.
$$\int \left(\frac{1}{x}+\frac{1}{x^n}\right)^n dx$$

16.
$$\int_0^\infty \sqrt[x]{x+1}\left(\frac{\ln(x+1)}{x^2}-\frac{1}{x(x+1)}\right)dx$$

17.
$$\int_0^1 \frac{\ln(x)+\frac{\ln^2(x)+\frac{\ln^3(x)+\ldots}{4}}{3}}{2}dx$$

18.
$$\int_0^{\frac{1}{2}} \frac{x}{x+\frac{x}{x+\ldots}}+x^2+x^3+x^4+\ldots\, dx$$

19.
$$\int_{-\infty}^{\infty} \frac{dx}{e^{x^2+x}+e^{x^2}}$$

20.
$$\int_0^7 \left(\sqrt[3]{\sqrt{x^2+1}+x}-\sqrt[3]{\sqrt{x^2+1}-x}\right)dx$$

21.
$$\int_0^{\frac{\pi}{4}} \left(3\sec(x)\tan(x)+\sec(x)\tan^3(x)\right)^4 dx$$

22.
$$\int \left(\sinh(x)\tan^{-1}(e^x)+\frac{1}{2}\right)dx$$

23.

$$\int e^{2x}\sqrt{e^{x^2}\sqrt[3]{e^{x^3}\sqrt[4]{e^{x^4}\sqrt[5]{e^{x^5}\ldots}}}}\,dx$$

24.

$$\int_1^e \frac{\ln(x)\ln(\sqrt{2}x)\ln(2x)}{x}dx$$

25.

$$\int \left(e^x - \frac{e^x}{x}\right)\left(e^{\frac{1}{x}} + \frac{e^{\frac{1}{x}}}{x}\right)dx$$

26.

$$\int \left(\frac{1}{(x+3)^2} - \frac{1}{(x+2)^2}\right)\frac{dx}{\sqrt{(x+2)(x+3)}}$$

27.

$$\int_1^{25} \left\lfloor \sqrt{x} \right\rfloor \bmod \left\lfloor x \right\rfloor dx$$

28.

$$\int_0^\infty e^{-x^{\frac{1}{3}}}x^{-\frac{1}{3}} + e^{-x^{\frac{2}{3}}}x^{-\frac{2}{3}}dx$$

29.

$$\int_1^e \left(\frac{d^2}{dx^2}\left[x^x\right]\right)\frac{dx}{x^x(1+\ln(x))}$$

30.

$$\int_0^{\sqrt{2}} \sec(\tan^{-1}(\cos(\sin^{-1}(x))))dx$$

31.

$$\int_1^{\sqrt{2}} \frac{dx}{1-\sqrt{1-x^{-2}}}$$

32.

$$\int_0^1 \frac{1}{x}\sqrt{\frac{x^\pi}{1-x^\pi}}dx$$

33.

$$\int_{-1}^1 \frac{\cos^{-1}(x)}{x^2+1}dx$$

34.

$$\int_0^{\frac{\pi}{2}} \csc^2 x \cot x - \frac{1}{x^3}dx$$

35.

$$\int_{\frac{\pi}{4}}^{\frac{3\pi}{4}} \sqrt{\sin(x)+\cos(x)+\sqrt{2}}\,dx$$

36.

$$\int_0^\infty \frac{1}{x} \ln\left(\frac{e^x+1}{2\cosh(x)}\right) dx$$

37.

$$\int_0^\infty x e^{-\left(x^2-\frac{1}{x^2}\right)^2} dx$$

38.

$$\int_0^\infty \frac{\cos(2x)-4\cos(x)+3}{x^3} dx$$

39.

$$\int_{-1}^1 \ln(\sinh^2(\ln(x)))dx$$

40.

$$\int (x+1)(x+\ln(x))\left(1+\frac{\ln(x)}{x}\right) dx$$

41.

$$\int_{-\infty}^0 \frac{2^x+\frac{1}{2}}{2^x+2^{-x}+1} dx$$

42.

$$\int_0^\infty \frac{(2x+5)}{(x+1)(x+2)(x+3)(x+4)} dx$$

43.

$$\int_{-1}^1 x^2 \tan^{-1}\left(\frac{e^x-x}{1+xe^x}\right) dx$$

44.

$$\int_0^{\frac{\pi}{2}} \frac{\pi}{2}\sin\left(\frac{\pi}{2}(1-\sin(x))\right)+\cos^{-1}\left(\frac{2}{\pi}\cos^{-1}\left(\frac{2}{\pi}x\right)\right) dx$$

45.

$$\int \ln\left(3x^5+5x^3\right)\left(\frac{x^2+1}{3x^3+5x}\right) dx$$

46.

$$\int_0^2 x\sqrt{x^{\ln(x)}\sqrt[3]{x^{\ln^2(x)}\sqrt[4]{x^{\ln^3(x)}\sqrt[5]{\cdots}}}}dx$$

47.

$$\int_0^{e^{2\pi}} \cos(\log(x))\log(x)dx$$

48.

$$\int_{-1}^\infty e^{-x^2}(5x^4-2x^6)dx$$

49.

$$\int_{-\frac{\pi}{3}}^{\frac{\pi}{3}} \frac{e^x \sec^2(x)}{\cosh(x)} dx$$

50.

$$\int_0^1 \frac{e^x\sqrt{1-x}}{e^x\sqrt{1-x}+\sqrt{ex}} dx$$

51.

$$\int_0^{\frac{\pi}{2}} \sqrt{\sin(x)\sqrt{\tan(x)\sqrt{\sin(x)\sqrt{\tan(x)\sqrt{\dots}}}}}\, dx$$

52.

$$\int_0^{\frac{\pi}{2}} \sqrt{\tan(x)+\cot(x)+2}\, dx$$

53.

$$\int \sqrt{\frac{e^x}{e^x-x}}\left(1-\frac{1}{x}\right) dx$$

54.

$$\int_1^e \ln\left(x^{x^x}\right)(x+\ln(x^x))dx$$

55.

$$\int_1^\infty \frac{\sqrt[x]{2}}{x^2+\sqrt[x]{2x^{2x}}} dx$$

56.

$$\int_{-1}^1 \frac{dx}{\sqrt{1+x}+\sqrt{1-x}+2}$$

Chapter 3

Problems made for Training

1. $$\int_{\frac{\pi}{3}}^{\frac{\pi}{2}} \frac{\tan(x)}{\sqrt{2+\tan^2(x)}}dx$$

2. $$\int (x^2+x)^2(x^2-x+1)^2dx$$

3. $$\int \frac{dx}{(x+1)\sqrt{x}+2x}$$

4. $$\int_{\frac{1}{4}}^{\frac{1}{2}} \frac{\sqrt{x-\sqrt{x-\frac{1}{4}}}}{\sqrt{x+\sqrt{x-\frac{1}{4}}}}dx$$

5. $$\int_1^{\sqrt{2}} (x^2+1)^3\left(x+\frac{1}{x^9}\right)dx$$

6. $$\int \cos(x)(x^3+6x)dx$$

7. $$\int e^{e^x+x}\left(\cos\left(e^{e^x}\right)+\cos(e^x)\right)dx$$

8. $$\int_0^{\ln(2)} \frac{\sec^{-1}(\cosh(x))}{\cosh(x)}dx$$

9. $$\int_0^{\infty} \frac{dx}{(1+x)(1+x^2)(1+x^4)}$$

10.

$$\int \sinh^5(x)e^{7x}dx$$

11.

$$\int_0^1 \frac{\tan(x)}{1-\tan(x)\tan(1-x)}dx$$

12.

$$\int_0^\infty \frac{\ln(x)(\ln(x)+1)}{x^{2x-1}}dx$$

13.

$$\int_{-4}^{-3} \frac{(x+4)^3}{(x+3)^2(x+5)^2+1}dx$$

14.

$$\int_0^\infty \sin\left(\frac{1}{x}\right)\sin\left(\frac{1}{3x}\right)dx$$

15.

$$\int_0^1 \frac{x-\sqrt{x-x^2}}{2x-1}dx$$

16.

$$\int x(\ln^2(x)+1)(\ln(x)+1)^2dx$$

17.

$$\int_0^2 \frac{3x^3}{2\sqrt{1+x^3}}+\frac{2x^2+2x}{3(x^2+2x)^{\frac{2}{3}}}dx$$

18.

$$\int_0^1 \frac{2x-1}{\sqrt{2x-x^2}-\sqrt{1-x^2}}dx$$

19.

$$\int_0^\infty xe^{-x}\operatorname{sech}(x)dx$$

Chapter 4

Solutions for Silver's Evil High School Integrals

4.1 There's no need to derive...

$$\int \frac{d}{dx}\left[\frac{x^2e^x}{\ln(x)}\right] \cdot \frac{\ln(x)}{x^2e^x}dx$$

There is actually no need to compute the derivative inside the integral. The derivative can actually be treated as du. Hence, let $u = \frac{x^2e^x}{\ln(x)}$:

$$= \int \frac{du}{u} = \ln|u| = \boxed{2\ln|x| + x - \ln|\ln|x|| + C}$$

Actually doing the tedious quotient rule is a self-torture horrible route.

4.2 The Power of Logs

$$\int x^{\frac{\ln(\ln(x))}{\ln(x)}} dx$$

We will use a sneaky log rule manipulation to alter and simplify the function. We shall perform the loophole log!

$$x^{\frac{\ln(\ln(x))}{\ln(x)}} = e^{\ln(x^{\frac{\ln(\ln(x))}{\ln(x)}})} = e^{\ln(x)\cdot\frac{\ln(\ln(x))}{\ln(x)}} = e^{\ln(\ln(x))} = \ln(x)$$

Hence by integration by parts:

$$= \int \ln(x)dx = \boxed{x\ln(x) - x + C}$$

4.3 An Exponential Staircase

$$\int \frac{e^{x+1+e^{x+1}}}{1+e^{1+e^{1+x}}}dx$$

Let's start small and easy, let $u = x + 1$:

$$= \int \frac{e^{u+e^u}}{1+e^{1+e^u}}du$$

Then let $w = e^{e^u}$:

$$= \int \frac{dw}{1+ew} = \frac{1}{e}\ln|1+ew| = \boxed{\frac{1}{e}\ln\left|1+e^{1+e^{x+1}}\right| + C}$$

4.4 A Fourth Root Trig Integral?

$$\int \sqrt[4]{\cos^4(x) - \cos(2x)}dx$$

Observe that: $\cos^4(x) - \cos(2x) = \cos^4(x) - 2\cos^2(x) + 1 = (1 - \cos^2(x))^2$.

$$= \int \sqrt{1-\cos^2(x)}dx = \int \sin(x)dx = \boxed{-\cos(x) + C}$$

4.5 An Intimidating Square

$$\int \left(\frac{\cos(x)}{\sin(x)+\cos(x)}\right)^2 dx$$

Note that $(\sin(x) + \cos(x))^2 = 1 + \sin(2x)$.

$$= \int \frac{\cos^2(x)}{1+\sin(2x)}dx = \frac{1}{2}\int \frac{1+\cos(2x)}{1+\sin(2x)}dx = \frac{1}{2}\int \sec^2(2x) - \sec(2x)\tan(2x) + \frac{\cos(2x)}{1+\sin(2x)}dx$$

$$= \boxed{\frac{1}{4}\left(\tan(2x) - \sec(2x) + \ln|1+\sin(2x)|\right) + C}$$

4.6 Double Tangent!

$$\int \tan(x)\tan(2x)dx$$

Note that

$$\tan(A+B) = \frac{\tan(A)+\tan(B)}{1-\tan(A)\tan(B)}.$$

Hence, $\tan(2x) = \frac{2\tan(x)}{1-\tan^2(x)}$.

$$= \int \frac{2\tan^2(x)}{1-\tan^2(x)}dx = 2\int \frac{1}{1-\tan^2(x)} - 1\ dx = 2\int \frac{\cos^2(x)}{\cos(2x)} - 1\ dx$$

$$= \int \frac{1+\cos(2x)}{\cos(2x)} - 2\ dx = \int \sec(2x) - 1\ dx$$

$$= \boxed{\frac{1}{2}\ln|\sec(2x)+\tan(2x)| - x + C}$$

4.7 Trig-Manipulation under the Root

$$\int \sqrt{1-2\sin(2x)+3\cos^2(x)}dx$$

Let $1 = \sin^2(x)+\cos^2(x)$, then

$$= \int \sqrt{\sin^2(x) - 4\sin(x)\cos(x) + 4\cos^2(x)}\ dx = \int (2\cos(x) - \sin(x))dx$$

$$= \boxed{2\sin(x)+\cos(x)+C}$$

4.8 An Awkward Mix of x and $\ln(x)$

$$\int \frac{\ln(x)-1}{x^2-\ln^2(x)}dx$$

$$= \int \frac{\ln(x)-1}{x^2} \cdot \frac{dx}{1-\left(\frac{\ln(x)}{x}\right)^2}$$

Let $u = \frac{\ln(x)}{x}$:

$$= \int \frac{du}{1-u^2} = \tanh^{-1}(u) = \boxed{\tanh^{-1}\left(\frac{\ln|x|}{x}\right) + C}$$

4.9 That 3 is what keeps all of us elementary...

$$\int \sqrt{\frac{x}{\ln(x)}}\left(1-\frac{1}{3\ln(x)}\right)dx$$

$$= \int \sqrt{\frac{x}{\ln(x)}} - \frac{\sqrt{x}}{3(\ln(x))^{\frac{3}{2}}}dx$$

We will perform semi-integration by parts. Let's only focus on the first term.

$$\int \sqrt{\frac{x}{\ln(x)}}dx = \frac{2x^{\frac{3}{2}}}{3\sqrt{\ln|x|}} + \frac{1}{3}\int \frac{\sqrt{x}}{(\ln(x))^{\frac{3}{2}}}dx$$

As you can see, that last term will cancel out with the second term from the original. Hence,

$$= \boxed{\frac{2}{3}x\sqrt{\frac{x}{\ln|x|}} + C}$$

4.10 A Very Evil Cosine Integral

$$\int \frac{\cos^2(x)+1}{\cos(x)\sqrt{\cos(x)}}dx$$

I find this to be the hardest integral out of all in this section.

Solution 1:

$$I = \int \sqrt{\cos(x)} + \sec^{3/2}(x)dx = \int \sqrt{\cos(x)}dx + \int \sec^2(x)\sqrt{\cos(x)}dx$$

Perform IBP on the second term.

$$I = \int \sqrt{\cos(x)}dx + \sqrt{\cos(x)}|\tan(x)| + \int \tan(x)\cdot\frac{\sin(x)}{2\sqrt{\cos(x)}}dx$$

$$I = \frac{|\sin(x)|}{\sqrt{\cos(x)}} + \int \sqrt{\cos(x)} + \frac{1-\cos^2(x)}{2\cos(x)\sqrt{\cos(x)}}dx$$

$$I = \frac{|\sin(x)|}{\sqrt{\cos(x)}} + \frac{1}{2}\int \sqrt{\cos(x)} + \sec^{3/2}(x)dx \Rightarrow I = \frac{|\sin(x)|}{\sqrt{\cos(x)}} + \frac{1}{2}I$$

$$I = \boxed{\frac{2|\sin(x)|}{\sqrt{\cos(x)}} + C}$$

Apparently, having the absolute value sign is necessary and the correct graph of the antiderivative.

Solution 2:

Let $u = \cos(x)$, then $\cos^{-1}(u) = x \to dx = -\frac{du}{\sqrt{1-u^2}}$:

$$= -\int \frac{u^2+1}{u\sqrt{u}} \cdot \frac{du}{\sqrt{1-u^2}} = -\int \frac{u^2+1}{u\sqrt{u-u^3}}du = -\int \frac{1+\frac{1}{u^2}}{\sqrt{\frac{1}{u}-u}}du$$

Let $w = \frac{1}{u} - u$, $dw = (-1 - \frac{1}{u^2})du$:

$$= \int \frac{dw}{\sqrt{w}} = 2\sqrt{w} = \boxed{2\sqrt{\sec(x) - \cos(x)} + C}$$

4.11 Multi-Sub!

$$\int \frac{dx}{\sqrt{x}\sqrt{1+\sqrt{x}}\sqrt{1+\sqrt{1+\sqrt{x}}}}$$

Let $u = \sqrt{1+\sqrt{x}}$, then $du = \frac{dx}{2\sqrt{1+\sqrt{x}}} \cdot \frac{1}{2\sqrt{x}}$.

$$= 4\int \frac{du}{\sqrt{1+u}} = 8\sqrt{1+u} = \boxed{8\sqrt{1+\sqrt{1+\sqrt{x}}} + C}$$

4.12 Why $(x+3)$?

$$\int (x+3)\sqrt{xe^x}dx$$

$$= \int x^{3/2}e^{x/2} + 3x^{1/2}e^{x/2}dx$$

If you look very closely, notice that

$$= 2\int x^{3/2} \cdot \frac{1}{2}e^{x/2} + \frac{3}{2}x^{1/2} \cdot e^{x/2}dx.$$

This is a reverse product rule, hence

$$= \boxed{2x\sqrt{xe^x} + C}$$

4.13 A Monstrous Looking Rational Polynomial

$$\int \left(\frac{(x+1)^{2019}(x^2+x+1)^{2019}}{(x^3+1)^{2019}}\right)^{2019} \left(\frac{(x-1)^{2019}(x^2-x+1)^{2019}}{(x^3-1)^{2019}}\right)^{2019} dx$$

$$= \int \left(\frac{(x+1)(x^2-x+1)(x-1)(x^2+x+1)}{(x^3+1)(x^3-1)}\right)^{2019^2} dx = \boxed{x+C}$$

4.14 Exponential Chain Tunnel

$$\int \left(\left(e^{e^{e^x}}+1\right)e^{e^x}+1\right)e^x dx$$

$$= \int e^{e^{e^x}} e^{e^x} e^x + e^{e^x} e^x + e^x \, dx$$

$$= \boxed{e^{e^{e^x}} + e^{e^x} + e^x + C}$$

4.15 A Temple of Secants and Tangents

$$\int \frac{\sec(x)+x+\tan(x)}{(\sec(x)+\tan(x))^2} dx$$

Note that $\frac{1}{\sec(x)+\tan(x)} = \sec(x) - \tan(x)$.

$$= \int (\sec(x)-\tan(x)) + x(\sec(x)-\tan(x))^2 dx$$

$$= \int (\sec(x)-\tan(x)) + x(2\sec^2(x) - 2\sec(x)\tan(x) - 1) dx$$

Perform IBP on the $x(2\sec^2(x) - 2\sec(x)\tan(x) - 1)$.

$$= 2x\tan(x) - 2x\sec(x) - x^2 + \int 3\sec(x) - 3\tan(x) + x \, dx$$

$$= \boxed{2x\tan(x) - 2x\sec(x) + 3\ln|\sec(x)+\tan(x)| - 3\ln|\sec(x)| - \frac{x^2}{2} + C}$$

4.16 A Random Exponential in the Denominator?

$$\int \frac{x}{x-1+e^{1-x}}dx$$

Multiply top and bottom by e^x:

$$= \int \frac{xe^x}{(x-1)e^x+e}dx$$

Let $u = (x-1)e^x + e$, then $du = xe^x dx$:

$$\int \frac{du}{u} = \ln|u| = \boxed{\ln|(x-1)e^x+e|+C}$$

4.17 Trig-Sub or U-Sub???

$$\int \frac{\ln(1+\sqrt{1-x^2})-\ln(x)}{x\sqrt{1-x^2}}dx$$

The fastest way is to let $u = \ln(1+\sqrt{1-x^2}) - \ln(x)$, but this isn't really intuitive. Most would let $x = \sin\theta$:

$$= \int \frac{\ln(1+\cos\theta)-\ln(\sin\theta)}{\sin\theta\cos\theta}\cos\theta d\theta$$

Note that $\frac{1+\cos\theta}{\sin\theta} = \cot(\theta/2)$.

$$= \int \ln(\cot(\theta/2))\csc\theta d\theta = -2\int \ln(\tan\theta)\csc(2\theta)d\theta$$

Let $u = \ln(\tan\theta)$, then $du = 2\csc(2\theta)d\theta$:

$$-\int u du = -\frac{u^2}{2} = \boxed{-\frac{(\ln(\tan(\sin^{-1}(x))))^2}{2}+C}$$

4.18 An Interesting Trig-Integral

$$\int \frac{\sin(\sin(x)-x)}{\cos(\sin(x))}dx$$

$$= \int \frac{\sin(\sin(x))\cos(x)-\cos(\sin(x))\sin(x)}{\cos(\sin(x))}dx = \int \tan(\sin(x))\cos(x)-\sin(x)dx$$

$$= \boxed{\ln|\sec(\sin(x))|+\cos(x)+C}$$

4.19 Algebra Manipulation like an Olympiad

$$\int \sqrt{x(x+1)(x+2)(x+3)+1}\,dx$$

$$= \int \sqrt{(x^2+3x)(x^2+3x+2)+1}\,dx$$

Let $a = x^2 + 3x$ to see things easier.

$$= \int \sqrt{a(a+2)+1}\,dx = \int \sqrt{a^2+2a+1}\,dx = \int (a+1)dx$$

$$\int (x^2+3x+1)dx = \boxed{\frac{1}{3}x^3 + \frac{3}{2}x^2 + x + C}$$

4.20 Just a Product Rule U-Sub

$$\int e^{e^x \cos(x)+x}(\sin(x) - \cos(x))dx$$

$$= \int e^{e^x \cos(x)} \cdot e^x(\sin(x) - \text{ccs}(x))dx$$

Let $u = e^x \cos(x)$, $du = e^x(\cos(x) - \sin(x))dx$:

$$= -\int e^u du = \boxed{-e^{e^x \cos(x)} + C}$$

4.21 What's with the $\sqrt{2}$?

$$\int \sqrt{1 + \sec^2(x) + 2\sqrt{2}\tan(x)}\,dx$$

$$= \int \sqrt{\tan^2(x) + 2\sqrt{2}\tan(x) + 2}\,dx = \int \tan(x) + \sqrt{2}\,dx$$

$$= \boxed{\ln|\sec(x)| + \sqrt{2}x + C}$$

4.22 We're gonna need to double the angle.

$$\int_0^1 \frac{\sqrt{1+x}}{\sqrt{1+\sqrt{1-x^2}}}dx$$

Let $x = \sin(2\theta)$:

$$= \int_0^{\frac{\pi}{4}} \frac{\sqrt{1+\sin(2\theta)}}{\sqrt{1+\cos(2\theta)}}2\cos(2\theta)d\theta = \int_0^{\frac{\pi}{4}} \frac{\sin\theta+\cos\theta}{\sqrt{2}\cos\theta} \cdot 2\cos(2\theta)d\theta$$

$$= \sqrt{2}\int_0^{\frac{\pi}{4}} \frac{\sin\theta(2\cos^2\theta-1)}{\cos\theta} + \cos(2\theta)d\theta = \sqrt{2}\int_{\frac{1}{\sqrt{2}}}^1 2u - \frac{1}{u}du + \frac{1}{\sqrt{2}}$$

$$\sqrt{2}\left[u^2 - \ln|u|\right]_{\frac{1}{\sqrt{2}}}^1 + \frac{1}{\sqrt{2}} = \boxed{\frac{1}{\sqrt{2}}(3-\ln(2))}$$

4.23 An Awkward Denominator

$$\int \frac{dx}{e^{-\sqrt{x}}+\sqrt{x}-1}$$

$$= \int \frac{e^{\sqrt{x}}}{1+e^{\sqrt{x}}(\sqrt{x}-1)}dx$$

Let $u = 1 + e^{\sqrt{x}}(\sqrt{x}-1)$, then $du = \frac{1}{2}e^{\sqrt{x}}dx$:

$$= \int \frac{2}{u}du = \boxed{2\ln|1+e^{\sqrt{x}}(\sqrt{x}-1)| + C}$$

4.24 An Ugly Integral Trick

$$\int_0^{\frac{\pi}{3}} \left(\frac{\sec(x)+\tan(x)}{\csc(x)+\cot(x)}\right)(\sec(x)+\csc(x))dx$$

Most people would just use Weierstrass substitution, however, there's an easy unintuitive way:

$$= \int_0^{\frac{\pi}{3}} e^{\ln(\sec(x)+\tan(x))-\ln(\csc(x)+\cot(x))}(\sec(x)+\csc(x))dx$$

Let $u = \ln(\sec(x)+\tan(x)) - \ln(\csc(x)+\cot(x))$, $du = \sec(x)+\csc(x)dx$:

$$= \int_{-\infty}^{\ln(1+\frac{2}{\sqrt{3}})} e^u du = \boxed{1+\frac{2}{\sqrt{3}}}$$

4.25 A Disgusting Looking Exponent

$$\int (x-2)e^{\left(\frac{1-x-xe^x}{e^x}\right)}dx$$

$$= \int (x-2)e^{e^{-x}-xe^{-x}-x}dx = \int (x-2)e^{-x}e^{e^{-x}-xe^{-x}}dx$$

Let $u = e^{-x} - xe^{-x}$, then $du = e^{-x}(x-2)dx$

$$= \int e^u du = \boxed{e^{e^{-x}-xe^{-x}} + C}$$

4.26 An Exponential Hiding in a Radical

$$\int \frac{x-1}{\sqrt{xe^x}}dx$$

$$= \int x^{1/2}e^{-x/2} - x^{-1/2}e^{-x/2}dx$$

$$= 2\int x^{1/2} \cdot \frac{1}{2}e^{-x/2} - \frac{1}{2}x^{-1/2}e^{-x/2}dx$$

As you can see, it is secretly reverse product rule. Hence,

$$= \boxed{2\sqrt{\frac{x}{e^x}} + C}$$

4.27 Factor out the Cosine!

$$\int e^{\sin(x)}e^{\cos(x)}\cos(2x)dx$$

$$= \int e^{(\sin(x)+\cos(x))}(\cos(x)+\sin(x))(\cos(x)-\sin(x))dx$$

Let $u = \sin(x) + \cos(x)$, then

$$= \int ue^u du = ue^u - e^u = \boxed{(\sin(x)+\cos(x))e^{\sin(x)+\cos(x)} - e^{\sin(x)+\cos(x)} + C}$$

4.28 Reciprocating the Secant-Tangent

$$\int \frac{\sec(x)\tan(x) - \sec^2(x)}{\sqrt{\sec(x)+\tan(x)}}dx$$

Note that $\sec(x) + \tan(x) = \frac{1}{\sec(x)-\tan(x)}$.

$$= \int (\sec(x)\tan(x) - \sec^2(x))\sqrt{\sec(x) - \tan(x)}\, dx$$

Let $u = \sec(x) - \tan(x)$, then $du = \sec(x)\tan(x) - \sec^2(x)dx$:

$$= \int \sqrt{u}du = \boxed{\frac{2}{3}(\sec(x) - \tan(x))^{3/2} + C}$$

4.29 A Crazy Secant Tri-Sub

$$\int_1^2 e^{\sec^{-1}(x)}\left(\frac{\sqrt{x^2-1}+1}{x}\right)dx$$

Let $x = \sec\theta$:

$$= \int_0^{\frac{\pi}{3}} e^{\theta} \cdot \frac{\tan\theta + 1}{\sec\theta} \cdot \sec\theta\tan\theta d\theta = \int_0^{\frac{\pi}{3}} e^{\theta}(\tan\theta + 1)\tan\theta d\theta$$

$$= \int_0^{\frac{\pi}{3}} e^{\theta}(\sec^2\theta + \tan\theta - 1)d\theta$$

The first two terms is a reverse product rule, hence we have

$$\left[e^{\theta}\tan\theta - e^{\theta}\right]_0^{\frac{\pi}{3}} = \boxed{e^{\frac{\pi}{3}}(\sqrt{3}-1)+1}$$

4.30 Sneakiest Trig-Manipulation

$$\int \frac{\tan(2x)(\sin^{14}(x) + \cos^{14}(x))}{(\sin^4(x)+\cos^4(x))(\sin^8(x)+\cos^8(x))}dx$$

Note: $\cos(2x) = \cos^4(x) - \sin^4(x)$

$$= \int \frac{\sin(2x)(\sin^{14}(x)+\cos^{14}(x))}{\cos^{16}(x) - \sin^{16}(x)}dx = 2\int \frac{\sin^{15}(x)\cos(x) + \cos^{15}(x)\sin(x)}{\cos^{16}(x) - \sin^{16}(x)}dx$$

Let $u = \cos^{16}(x) - \sin^{16}(x)$, then $du = -16\cos^{15}(x)\sin(x) - 16\sin^{15}(x)\cos(x)dx$:

$$= -\frac{1}{8}\int \frac{du}{u} = \boxed{-\frac{1}{8}\ln|\cos^{16}(x) - \sin^{16}(x)| + C}$$

4.31 A Suspicious "$-x$"

$$\int \frac{e^{e^x}+1}{e^{e^x-x}+1}dx$$

$$= \int \frac{e^{e^x}e^x + e^x}{e^{e^x}+e^x}dx = \boxed{\ln|e^{e^x}+e^x| + C}$$

4.32 A Rude Way to Hide U-Substitution

$$\int \left(\frac{\cos(x)}{\cos(x+\sin(x)\cos(x))}\right)^2 dx$$

$$= \int \sec^2(x+\sin(x)\cos(x))\cos^2(x)dx$$

Let $u = x + \frac{1}{2}\sin(2x)$, $du = 1+\cos(2x) = 2\cos^2(x)dx$:

$$= \frac{1}{2}\int \sec^2(u)du = \boxed{\frac{1}{2}\tan(x+\sin(x)\cos(x)) + C}$$

4.33 Deconstruct the Cosine

$$\int \frac{\cos(x)}{\sqrt{2}\cos(\frac{x}{2})+1}dx$$

$$= \int \frac{2\cos^2(\frac{x}{2})-1}{\sqrt{2}\cos(\frac{x}{2})+1}dx = \int \sqrt{2}\cos\left(\frac{x}{2}\right) - 1\ dx = \boxed{2\sqrt{2}\sin\left(\frac{x}{2}\right) - x + C}$$

4.34 A Sneaky Easy Arcsine Integral

$$\int \frac{1-x}{\sqrt{e^{2x}-x^2}}dx$$

Multiply e^{-x} top and bottom and also put it inside the radical.

$$= \int \frac{(1-x)e^{-x}}{\sqrt{1-(xe^{-x})^2}}dx$$

Let $u = xe^{-x}$, then $du = e^{-x}(1-x)dx$:

$$= \int \frac{du}{\sqrt{1-u^2}} = \sin^{-1}(u) = \boxed{\sin^{-1}(xe^{-x}) + C}$$

4.35 An Unusual Arcsecant Exponential Integral

$$\int \frac{1}{x^2}e^{\sec^{-1}(x)}dx$$

Let $x = \sec\theta$:

$$= \int e^{\theta}\sin\theta d\theta = \frac{1}{2}e^{\theta}(\sin\theta - \cos\theta) = \boxed{\frac{e^{\sec^{-1}(x)}}{2}\left(\frac{\sqrt{x^2-1}-1}{x}\right) + C}$$

4.36 The Pyramid of Radicals!!!

$$\int \frac{1+\frac{1+\frac{1+\frac{1}{2\sqrt{x}}}{2\sqrt{x+\sqrt{x}}}}{2\sqrt{x+\sqrt{x+\sqrt{x}}}}}{2\sqrt{x+\sqrt{x+\sqrt{x+\sqrt{x}}}}}dx$$

Clearly, this is a huge multi u-substitution. Let $u = \sqrt{x+\sqrt{x+\sqrt{x+\sqrt{x}}}}$. By testing this derivative, you will notice that this substitution is literally the answer to this integral.

$$\boxed{\sqrt{x+\sqrt{x+\sqrt{x+\sqrt{x}}}} + C}$$

4.37 A Challenging Infinite Function

$$\int_2^6 \sqrt{x - \frac{1}{x - \frac{1}{x - \dots}}}dx$$

$$y = x - \frac{1}{x - \frac{1}{x - \dots}} = x - \frac{1}{y} \Rightarrow y^2 - xy + 1 = 0 \Rightarrow y = \frac{x \pm \sqrt{x^2 - 4}}{2}$$

$$= \int_2^6 \sqrt{\frac{x + \sqrt{x^2 - 4}}{2}}\, dx$$

Let $x = 2\sec\theta$:

$$= \frac{1}{\sqrt{2}} \int_0^{\sec^{-1}(3)} \sqrt{2\sec\theta + 2\tan\theta}\, 2\sec\theta\tan\theta d\theta = 2\int_0^{\sec^{-1}(3)} \sqrt{\sec\theta + \tan\theta} \sec\theta\tan\theta d\theta$$

Let $u = \sec\theta + \tan\theta$, $du = u\sec\theta d\theta$, and $\frac{1}{u} = \sec\theta - \tan\theta d\theta$. Then $u - \frac{1}{u} = 2\tan\theta$.

$$= \int_1^{3+2\sqrt{2}} \frac{1}{\sqrt{u}}\left(u - \frac{1}{u}\right) du = \int_1^{(1+\sqrt{2})^2} \sqrt{u} - u^{-3/2} du = \left[\frac{2}{3}u^{3/2} + 2u^{-1/2}\right]_1^{(1+\sqrt{2})^2}$$

$$= \boxed{\frac{16\sqrt{2}}{3}}$$

4.38 An Interesting Tetration

$$\int x^x \left(\ln^2(x) + \ln(x) + \frac{1}{x}\right) dx$$

$$= \int x^x(1 + \ln(x))\ln(x) + x^x \cdot \frac{1}{x} dx$$

This is secretly reverse product rule since the derivative of x^x is $x^x(1 + \ln(x))$. Hence,

$$= \boxed{x^x \ln(x) + C}$$

4.39 A Beautiful Algebra Manipulation

$$\int \frac{x^x}{x^x+\sqrt{x}} + \frac{x^{1-x}}{x^{1-x}+\sqrt{x}} dx$$

Observe that

$$\frac{x^{1-x}}{x^{1-x}+\sqrt{x}} = \frac{x}{x+x^x\sqrt{x}} = \frac{\sqrt{x}}{\sqrt{x}+x^x}.$$

Then,

$$\int \frac{x^x}{x^x+\sqrt{x}} + \frac{\sqrt{x}}{x^x+\sqrt{x}} dx = \boxed{x+C}$$

4.40 Some Rearrangements

$$\int x^e(e^x+e^{-x}) + x^{e-1}(e^{x+1}-e^{1-x})dx$$

$$= \int x^e \cdot e^x + e^x \cdot ex^{e-1} + x^e e^{-x} - ex^{e-1}e^{-x} dx$$

This is secretly a double reverse product rule!!!

$$= \boxed{x^e e^x + x^e e^{-x} + C}$$

4.41 A Nice Nested Radical Simplification

$$\int \sqrt{1+x\sqrt{1+(x+1)\sqrt{1+(x+2)(x+4)}}}\, dx$$

$$= \int \sqrt{1+x\sqrt{1+(x+1)(x+3)}}\, dx = \int \sqrt{1+x(x+2)}\, dx = \int (x+1)dx = \boxed{\frac{x^2}{2}+x+C}$$

4.42 Diabolically Mesmerizing

$$\int \left(\frac{\cos(x)}{\sqrt{\sin(x)}} - \frac{\sin(x)}{\sqrt{\cos(x)}}\right)\left(\frac{\sin(x)-\cos(x)}{\sqrt{\sin(x)}-\sqrt{\cos(x)}}\right) dx$$

Notice that

$$\frac{\sin(x)-\cos(x)}{\sqrt{\sin(x)}-\sqrt{\cos(x)}} = \sqrt{\sin(x)} + \sqrt{\cos(x)}.$$

Then

$$= \int \left(\frac{\cos(x)}{\sqrt{\sin(x)}} - \frac{\sin(x)}{\sqrt{\cos(x)}}\right)\left(\sqrt{\sin(x)} + \sqrt{\cos(x)}\right) dx$$

$$= \int \cos(x) - \sin(x) + \frac{\cos(x)\sqrt{\cos(x)}}{\sqrt{\sin(x)}} - \frac{\sin(x)\sqrt{\sin(x)}}{\sqrt{\cos(x)}} dx$$

$$= \sin(x) + \cos(x) + \int \frac{\sqrt{2}\cos(2x)}{\sqrt{\sin(2x)}} dx = \boxed{\sin(x) + \cos(x) + \sqrt{2\sin(2x)} + C}$$

4.43 Easy Decomposition

$$\int \frac{(x+1)\sin(x) + (x-1)\cos(x)}{x(\sin(x)-\cos(x))} dx$$

$$= \int \frac{x(\sin(x)+\cos(x)) + \sin(x) - \cos(x)}{x(\sin(x)-\cos(x))} dx$$

$$= \int \frac{\sin(x)+\cos(x)}{\sin(x)-\cos(x)} + \frac{1}{x} dx = \boxed{\ln|\sin(x)-\cos(x)| + \ln|x| + C}$$

4.44 IBP with Bravery

$$\int \tan^{-1}(\sec(x) + \tan(x)) dx$$

Notice that

$$\frac{d}{dx}\left[\tan^{-1}(\sec(x)+\tan(x))\right] = \frac{\sec(x)(\sec(x)+\tan(x))}{1+(\sec(x)+\tan(x))^2} = \frac{\sec^2(x) + \sec(x)\tan(x)}{2\sec^2(x) + 2\sec(x)\tan(x)} = \frac{1}{2}.$$

Then by IBP,

$$= x\tan^{-1}(\sec(x)+\tan(x)) - \frac{1}{2}\int x dx = \boxed{x\tan^{-1}(\sec(x)+\tan(x)) - \frac{x^2}{4} + C}$$

4.45 A Suspicious $2x$

$$\int \frac{\sin^2(x)}{2x - \sin(2x)} dx$$

$$= \frac{1}{2} \int \frac{1 - \cos(2x)}{2x - \sin(2x)} dx$$

Let $u = 2x - \sin(2x)$, then $du = 2(1 - \cos(2x))dx$:

$$= \frac{1}{4} \int \frac{du}{u} = \boxed{\frac{1}{4} \ln|2x - \sin(2x)| + C}$$

4.46 IBP Love Triangle

$$\int \sec^2(x) \tan(x) \ln(\sec(x)) dx$$

Perform IBP by deriving $\ln(\sec(x))$ and integrating $\tan(x)\sec^2(x)$.

$$= \frac{1}{2} \tan^2(x) \ln|\sec(x)| - \frac{1}{2} \int \tan^3(x) dx$$

$$= \frac{1}{2} \tan^2(x) \ln|\sec(x)| - \frac{1}{2} \int \tan(x)(\sec^2(x) - 1) dx$$

$$= \boxed{\frac{1}{2} \tan^2(x) \ln|\sec(x)| - \frac{1}{4} \tan^2(x) + \frac{1}{2} \ln|\sec(x)| + C}$$

4.47 A Random Exponential Function in the Radical

$$\int \frac{x+1}{x\sqrt{xe^x - 1}} dx$$

Multiply top and bottom by e^x:

$$= \int \frac{e^x(x+1)}{xe^x\sqrt{xe^x - 1}} dx$$

Let $u = xe^x$:

$$= \int \frac{du}{u\sqrt{u-1}} = 2\tan^{-1}(\sqrt{u-1}) = \boxed{2\tan^{-1}(\sqrt{xe^x - 1}) + C}$$

4.48 Tangent Twinsies???

$$\int \frac{x}{x\tan(x)+1} + \frac{1}{\sqrt{x}\tan(\sqrt{x})+1}dx$$

Observe that

$$= \int \frac{x}{x\tan(x)+1}dx = \int \frac{x\cos(x)}{x\sin(x)+\cos(x)}dx = \ln|x\sin(x) + \cos(x)| + C$$

$$= \int \frac{dx}{\sqrt{x}\tan(\sqrt{x})+1} = \int \frac{2u}{u\tan(u)+1}du = 2\ln|\sqrt{x}\sin(\sqrt{x}) + \cos(\sqrt{x})| + C$$

Hence,

$$= \boxed{\ln|x\sin(x)+\cos(x)| + 2\ln|\sqrt{x}\sin(\sqrt{x})+\cos(\sqrt{x})| + C}$$

4.49 One Adjustment can Really Throw You Off

$$\int \sqrt{x}\ln(x)\sqrt{\ln(x)-1}\ dx$$

$$= \int \sqrt{x\ln(x) - x}\ln(x)dx$$

Let $u = x\ln(x) - x, du = \ln(x)dx$:

$$= \int \sqrt{u}du = \boxed{\frac{2}{3}(x\ln|x| - x)^{3/2} + C}$$

4.50 Don't you Dare Factor It

$$\int \frac{\tan(x)}{\sqrt{\sec^4(x)-1}}dx$$

Multiply top and bottom by $\sec(x)$.

$$= \int \frac{\sec(x)\tan(x)}{\sec(x)\sqrt{\sec^4(x)-1}}dx = \int \frac{du}{u\sqrt{u^4-1}}$$

After $u = \sec(x)$, let $w = u^2$:

$$= \frac{1}{2}\int \frac{dw}{w\sqrt{w^2-1}} = \frac{1}{2}\sec^{-1}(w) = \boxed{\frac{1}{2}\sec^{-1}(\sec^2(x)) + C}$$

4.51 Two Terms are Elementary Together

$$\int \frac{\ln(e^x - 1) - x}{e^x - 1} dx$$

Let $u = \ln(e^x - 1) - x$, then $du = \frac{dx}{e^x - 1}$:

$$= \int u du = \boxed{\frac{1}{2}(\ln(e^x - 1) - x)^2 + C}$$

4.52 A Party in the Log

$$\int \frac{\ln((x+1)^{(x+1)} x^x)}{x(x+1)} dx$$

$$= \int \frac{(x+1)\ln(x+1) + x\ln(x)}{x(x+1)} dx = \int \frac{\ln(1+x)}{x} + \frac{\ln(x)}{x+1} dx$$

By reverse product rule,

$$= \boxed{\ln|x| \ln|x+1| + C}$$

4.53 Experimenting with Trig-Identities

$$\int \frac{e^{\sec(x) - \tan(x)}}{1 + \sin(x)} dx$$

Observe that

$$= \frac{1}{1 + \sin(x)} = \frac{1 - \sin(x)}{\cos^2(x)} = \sec^2(x) - \sec(x)\tan(x).$$

$$= \int e^{\sec(x) - \tan(x)}(\sec^2(x) - \sec(x)\tan(x)) dx = \boxed{-e^{\sec(x) - \tan(x)} + C}$$

4.54 An Awkward Exponent of Square Roots

$$\int \left(1-\frac{1}{x^2}\right) e^{\sqrt{x}+\frac{1}{\sqrt{x}}} dx$$

Let $u = \sqrt{x} + \frac{1}{\sqrt{x}}$, then

$$\begin{aligned} du &= \frac{1}{2\sqrt{x}} - \frac{1}{2x\sqrt{x}} dx \\ du &= \frac{1}{2\sqrt{x}}\left(1-\frac{1}{x}\right) dx \\ udu &= \frac{1}{2\sqrt{x}}\left(\sqrt{x}+\frac{1}{\sqrt{x}}\right)\left(1-\frac{1}{x}\right) dx \\ udu &= \frac{1}{2}\left(1-\frac{1}{x^2}\right) dx. \end{aligned}$$

Hence,

$$= \int 2ue^u du = 2(ue^u - e^u) = \boxed{2\left(\sqrt{x}+\frac{1}{\sqrt{x}}\right) e^{\sqrt{x}+\frac{1}{\sqrt{x}}} - 2e^{\sqrt{x}+\frac{1}{\sqrt{x}}} + C}$$

4.55 A Secret Hidden Manipulation

$$\int_{-\frac{\pi}{4}}^{\frac{\pi}{4}} \frac{\sqrt{\sec(2x)}}{\cos(x)+\sqrt{\cos(2x)}} dx$$

$$= 2\int_0^{\frac{\pi}{4}} \frac{\sqrt{\sec(2x)}}{\cos(x)+\sqrt{2\cos^2(x)-1}} \cdot \frac{\sqrt{\cos(2x)}-\cos(x)}{\sqrt{2\cos^2(x)-1}-\cos(x)} dx$$

$$= 2\int_0^{\frac{\pi}{4}} \frac{1-\cos(x)\sqrt{\sec(2x)}}{\cos^2(x)-1} dx = 2\int_0^{\frac{\pi}{4}} \frac{\cos(x)\sqrt{\sec(2x)}-1}{\sin^2(x)} dx$$

$$= 2\int_0^{\frac{\pi}{4}} \frac{\cos(x)}{\sin^2(x)\sqrt{1-2\sin^2(x)}} - \csc^2(x) dx = 2 - \lim_{x\to 0} 2\cot(x) + 2\int_0^{\frac{1}{\sqrt{2}}} \frac{du}{u^2\sqrt{1-2u^2}}$$

Let $u = \frac{\sin\theta}{\sqrt{2}}$:

$$= 2 - \lim_{x\to 0}\cot(x) + \sqrt{2}\int_0^{\frac{\pi}{2}} \frac{2d\theta}{\sin^2\theta} = 2 - 2\lim_{x\to 0}\left(\cot(x) - \frac{\sqrt{1-2x^2}}{x}\right) = \boxed{2}$$

Note: A very challenging limit...

$$\lim_{x\to 0}\left(\cot(x) - \frac{\sqrt{1-2x^2}}{x}\right) = 0.$$

4.56 A Tricky Trig-Sub

$$\int_0^1 \frac{x\sqrt{\sqrt{1-x}+\sqrt{1+x}}}{\sqrt{1-x^2}}dx$$

Let $x = \cos(2\theta)$:

$$= \int_0^{\frac{\pi}{4}} \frac{\cos(2\theta)\sqrt{\sqrt{2}(\sin\theta+\cos\theta)}}{\sin(2\theta)} \cdot 2\sin(2\theta)d\theta$$

$$= 2\sqrt[4]{2}\int_0^{\frac{\pi}{4}} (\cos\theta-\sin\theta)(\cos\theta+\sin\theta)^{3/2}d\theta$$

Let $u = \sin\theta+\cos\theta$, $du = (\cos\theta-\sin\theta)d\theta$:

$$= 2\sqrt[4]{2}\int_1^{\sqrt{2}} u^{3/2}du = 2^{5/4}\left[\frac{2}{5}u^{5/2}\right]_1^{\sqrt{2}} = \frac{2^{9/4}}{5}\left(2^{5/4}-1\right) = \boxed{\frac{2^{7/2}-2^{9/4}}{5}}$$

4.57 Getting Out of the Seventh Power

$$\int (256^x + 2^{x+1})^7 dx$$

$$= \int (2^{8x}+2^{x+1})^7dx = \int 2^{7x}(2^{7x}+2)^7dx$$

Let $u = 2^{7x}+2$, then

$$= \frac{1}{\ln(2)}\int u^7du = \boxed{\frac{(2^{7x}+2)^8}{8\ln(2)}+C}$$

4.58 There's Gold Hidden Underneath

$$\int_0^1 \frac{2^x}{(\sqrt{5}+1)^x+(\sqrt{5}+3)^x}dx$$

Divide top and bottom by 2^x:

$$= \int_0^1 \frac{dx}{\varphi^x+(1+\varphi)^x} = \int_0^1 \frac{\varphi^{-x}}{1+(1+\frac{1}{\varphi})^x}dx$$

Note that $1+\frac{1}{\varphi} = \varphi$. Let $u = \varphi^{-x}$:

$$= \int_0^1 \frac{\varphi^{-x}}{1+\varphi^x}dx = \int_0^1 \frac{\varphi^{-2x}}{\varphi^{-x}+1}dx = \frac{1}{\ln(\varphi)}\int_{\frac{1}{\varphi}}^1 1-\frac{1}{u+1}du = \boxed{\frac{1-\frac{1}{\varphi}-\ln(2)+\ln(\varphi)}{\ln(\varphi)}}$$

4.59 A Sneaky Secant Maneuver

$$\int_0^{\frac{\pi}{3}} \frac{1 + x\tan(x)}{\sqrt{x\cos(x)}}dx$$

Multiply top and bottom by $\sec(x)$. Let $u = x\sec(x)$, then $du = \sec(x) + x\sec(x)\tan(x)dx$:

$$= \int_0^{\frac{\pi}{3}} \frac{\sec(x) + x\sec(x)\tan(x)}{\sqrt{x\sec(x)}}dx = \int_0^{\frac{2\pi}{3}} \frac{du}{\sqrt{u}} = \boxed{2\sqrt{\frac{2\pi}{3}}}$$

4.60 Fear of Partial Fractions

$$\int \frac{x^2}{(x^2 - 2x + 4)(x^2 + 3x + 9)(x^2 - x - 6)}dx$$

Notice that $(x^2 - x - 6) = (x - 3)(x + 2)$. Then we secretly have:

$$\int \frac{x^2}{(x^3 - 27)(x^3 + 8)}dx = \frac{1}{3}\int \frac{1}{(u - 35)u}du = \frac{1}{3}\int \frac{-\frac{1}{35}}{u} + \frac{\frac{1}{35}}{u - 35}du$$

$$= \frac{1}{105}\ln|u - 35| - \frac{1}{105}\ln|u|$$

$$= \boxed{\frac{1}{105}\ln|x^3 - 35| - \frac{1}{105}\ln|x^3| + C}$$

4.61 No Other Choice...

$$\int \frac{\sqrt{\cos(2x)}}{\sin(x)}dx$$

$$= \int \frac{\sin(x)\sqrt{2\cos^2(x) - 1}}{1 - \cos^2(x)}dx = -\int \frac{\sqrt{2u^2 - 1}}{1 - u^2}du$$

Let $u = \frac{1}{\sqrt{2}}\sec\theta$, then $du = \frac{1}{\sqrt{2}}\sec\theta\tan\theta d\theta$:

$$= -\sqrt{2}\int \frac{\sec\theta\tan^2\theta}{2 - \sec^2\theta}d\theta = -\sqrt{2}\int \frac{\sin^2\theta}{(2\cos^2\theta - 1)\cos\theta}d\theta = -\sqrt{2}\int \frac{(1 - \cos^2\theta) - \cos^2\theta + \cos^2\theta}{(2\cos^2\theta - 1)\cos\theta}d\theta$$

$$= \sqrt{2}\int \sec\theta - \frac{\cos\theta}{1 - 2\sin^2\theta}d\theta = \sqrt{2}\ln|\sec\theta + \tan\theta| - \frac{1}{\sqrt{2}}\int \frac{dw}{\frac{1}{2} - w^2}$$

$$= \sqrt{2}\ln|\sec\theta + \tan\theta| - \tanh^{-1}(\sqrt{2}\sin\theta) + C$$

$$= \boxed{\sqrt{2}\ln|\sqrt{2}\cos(x) + \sqrt{\cos(2x)}| - \tanh^{-1}(\sqrt{2}\sin(\sec^{-1}(\sqrt{2}\cos(x)))) + C}$$

4.62 A Disgusting Arctangent Function

$$\int \frac{2x^3 - 1}{x^2 + (x^3 + 1)^2} dx$$

Divide top and bottom by x^2:

$$= \int \frac{2x - \frac{1}{x^2}}{1 + (x^2 + \frac{1}{x})^2} dx$$

Let $u = x^2 + \frac{1}{x}$:

$$= \int \frac{du}{1 + u^2} = \tan^{-1}(u) = \boxed{\tan^{-1}\left(x^2 + \frac{1}{x}\right) + C}$$

Chapter 5

Solutions for Silver's University Competition Problems

5.1 A Nice Trig Cancellation

$$\int \cos(\sin(x) - x) + \cos(\sin(x) + x)dx$$

$$= \int \cos(\sin(x))\cos(x) + \sin(\sin(x))\sin(x) + \cos(\sin(x))\cos(x) - \sin(\sin(x))\sin(x)dx$$

$$= \int 2\cos(\sin(x))\cos(x)dx = \boxed{2\sin(\sin(x)) + C}$$

5.2 An Odd Log Tower

$$\int x^{\frac{x^{\frac{x}{\ln(x)}}}{\ln(x)} + \frac{x}{\ln(x)}} dx$$

Observe that

$$x^{\frac{x}{\ln(x)}} = e^{\ln(x^{\frac{x}{\ln(x)}})} = e^{\ln(x)\cdot\frac{x}{\ln(x)}} = e^x.$$

$$= \int x^{\frac{e^x}{\ln(x)}} e^x dx = \int e^{e^x} e^x dx = \boxed{e^{e^x} + C}$$

5.3 A Brutal Glasser Trig-Integral

$$\int_{-\infty}^{\infty} \frac{\sin(2x)\cos(\frac{1}{x}) - \sin(\frac{1}{x})\cos(2x)}{2x^3 - x} dx$$

By trig-identity:

$$= \int_{-\infty}^{\infty} \frac{\sin(2x - \frac{1}{x})}{x(2x^2 - 1)} dx = 2\int_{0}^{\infty} \frac{\sin(2x - \frac{1}{x})}{x^2(2x - \frac{1}{x})} dx = 2\int_{0}^{\infty} \frac{\sin(x - \frac{2}{x})}{(x - \frac{2}{x})} dx$$

By Glasser's Master Theorem,

$$= 2\int_{0}^{\infty} \frac{\sin(x)}{x} dx = \boxed{\pi}$$

5.4 A Sneaky Unusual Factor

$$\int \left(e^{x^3 - x}\right)(3x^3 - x + 1)\, dx$$

Take out the x from the right-side parenthesis: $x(3x^2 - 1 + \frac{1}{x})$. Note that $x = e^{\ln(x)}$. So,

$$\int \left(e^{x^3 - x + \ln(x)}\right)\left(3x^2 - 1 + \frac{1}{x}\right) dx = \int e^u du = \boxed{e^{x^3 - x + \ln(x)} + C}$$

5.5 A Cool Glasser's Exponential Integral

$$\int_{-\infty}^{\infty} \frac{e^{x + \frac{1}{x}}}{e^{\frac{2}{x}} + e^{2x}} dx$$

Using Glasser's Master Theorem:

$$\int_{-\infty}^{\infty} \frac{e^{x - \frac{1}{x}} \cdot e^{\frac{2}{x}}}{e^{\frac{2}{x}} + e^{2x}} dx = \int_{-\infty}^{\infty} \frac{e^{x - \frac{1}{x}}}{1 + e^{2(x - \frac{1}{x})}} dx = \int_{-\infty}^{\infty} \frac{e^{x}}{1 + e^{2x}} dx = \boxed{\frac{\pi}{2}}$$

5.6 A Small Trig Integral yet Tricky

$$\int \frac{dx}{\cos x \sqrt{\sin x}}$$

Rewrite the integrand and let $u = \sqrt{\sin(x)}$:

$$= \int \frac{2\cos x\, dx}{2\sqrt{\sin x}} \frac{1}{\cos^2 x} = \int \left(\frac{1}{1 - u^2} + \frac{1}{1 + u^2}\right) du = \boxed{\tanh^{-1}\sqrt{\sin x} + \tan^{-1}\sqrt{\sin x}}$$

5.7 An Intimidating Arcsecant Integral

$$\int_1^{\sqrt{2}} \frac{2x(\sqrt{x^2-1}+\sec^{-1}(x))}{(x^2+1)^2}dx$$

Integrate by parts by deriving $\sqrt{x^2-1}+\sec^{-1}(x)$ and integrating $\frac{2x}{(x^2+1)^2}$.

$$= -\frac{1+\frac{\pi}{4}}{3} + \int_1^{\sqrt{2}} \frac{x}{(1+x^2)\sqrt{x^2-1}} + \frac{1}{x(x^2+1)\sqrt{x^2-1}}dx$$

$$= -\frac{1+\frac{\pi}{4}}{3} + \int_1^{\sqrt{2}} \frac{1}{(1+x^2)\sqrt{x^2-1}}\left(\frac{x^2+1}{x}\right)dx = -\frac{1+\frac{\pi}{4}}{3} + \int_1^{\sqrt{2}} \frac{dx}{x\sqrt{x^2-1}}$$

$$= -\frac{1}{3} - \frac{\pi}{12} + \frac{\pi}{4} = \boxed{\frac{\pi}{6} - \frac{1}{3}}$$

5.8 It Looks Simple...But it Isn't...

$$\int_2^3 \frac{dx}{\sqrt{1-\frac{2}{x}}}$$

$$\int_2^3 \frac{dx}{\sqrt{1-\frac{2}{x}}} = \int_2^3 \frac{\sqrt{x}}{\sqrt{x-2}}dx$$

Let $x = 2\sec^2\theta$:

$$\int_1^{\sqrt{\frac{3}{2}}} 4\sec^3\theta d\theta = \boxed{\sqrt{3}+\ln(2+\sqrt{3})}$$

5.9 A Surprisingly Satisfying IBP

$$\int_0^{\frac{\pi}{4}} \ln(\sec x + \tan x)\left(\cos^2 x - \frac{1}{2}\right)dx$$

Derive $\ln(\sec x + \tan x)$ and integrate $\cos^2 x - \frac{1}{2}$.

$$= \left[\frac{1}{4}\ln(\sec x + \tan x)\sin(2x)\right]_0^{\frac{\pi}{4}} - \frac{1}{4}\int_0^{\frac{\pi}{4}} \sec(x)\sin(2x)dx$$

$$= \frac{\ln(1+\sqrt{2})}{4} - \frac{1}{2}\int_0^{\frac{\pi}{4}} \sin(x)dx = \boxed{\frac{\ln(1+\sqrt{2})}{4} + \frac{1}{2\sqrt{2}} - \frac{1}{2}}$$

5.10 An Unexpected Weierstrass

$$\int \frac{dx}{\sqrt{\sin(x)(\cos(x)+1)}}$$

Let $u = \tan\left(\frac{x}{2}\right)$,

$$\int \frac{2du}{(1+u^2)\sqrt{\left(\frac{2u}{1+u^2}\right)\left(\frac{1-u^2}{1+u^2}+\frac{1+u^2}{1+u^2}\right)}} = \int \frac{2(1+u^2)du}{(1+u^2)\sqrt{2u(2)}} = \int \frac{du}{\sqrt{u}} = \boxed{2\sqrt{\tan\left(\frac{x}{2}\right)}+C}$$

5.11 Forcing Weierstrass

$$\int \frac{dx}{\sin x\left(\ln(\sin x) - \ln(1+\cos x)\right)}$$

$$\int \frac{dx}{\sin(x)\ln\left(\tan\left(\frac{x}{2}\right)\right)} = \int \frac{\sec^2\left(\frac{x}{2}\right)}{2\tan\left(\frac{x}{2}\right)\ln\left(\tan\left(\frac{x}{2}\right)\right)}dx = \int \frac{1}{2u\ln(u)}du$$

$$= \boxed{\frac{1}{2}\ln\left|\ln\left|\tan\left(\frac{x}{2}\right)\right|\right|+C}$$

5.12 An Infinite Void Inside the Logarithm

$$\int_2^e \ln\left(\sqrt[x]{e\sqrt[x]{e\sqrt[x]{e\sqrt[x]{\ldots}}}}\right)dx$$

Simplifying the infinite function gives us

$$\ln\left(\sqrt[x]{e\sqrt[x]{e\sqrt[x]{e\sqrt[x]{\ldots}}}}\right) = \frac{1+\frac{1+\frac{1+\ldots}{x}}{x}}{x}$$

Notice that

$$w = \frac{1+w}{x} \to wx = w+1 \to w = \frac{1}{x-1}$$

Hence,

$$= \int_2^e \frac{1}{x-1}dx = \boxed{\ln(e-1)}$$

5.13 A Deadly Trig-Integral

$$\int \frac{\tan(x)}{\sqrt{\cos(2x)}}dx$$

If you get $\sin^{-1}(\frac{\sec(x)}{\sqrt{2}})$, you are incorrect. This answer is unacceptable.

Let $u = \cos(2x)$, observe that

$$\begin{aligned} du &= -2\sin(2x)dx \\ \frac{du}{1+u} &= \frac{-2\sin(2x)}{1+\cos(2x)}dx \\ &= -2\tan(x)dx. \end{aligned}$$

$$-\frac{1}{2}\int \frac{du}{\sqrt{u}(u+1)} = \boxed{-\tan^{-1}\left(\sqrt{\cos(2x)}\right) + C}$$

5.14 A Terrifying Polynomial

$$\int_0^1 (x^{2022} + x)^{2020}dx$$

$$= \int_0^1 x^{2020}(x^{2021}+1)^{2020}dx = \frac{1}{2021}\int_1^2 u^{2020}du = \boxed{\frac{2^{2021}-1}{2021^2}}$$

5.15 A Sneaky U-Sub

$$\int \left(\frac{1}{x} + \frac{1}{x^n}\right)^n dx$$

This is the same concept as the previous integral.

$$= \int \frac{1}{x^n}\left(1 + \frac{1}{x^{n-1}}\right)^n dx$$

Let $u = 1 + \frac{1}{x^{n-1}}$,

$$= \int (1-n)u^n du = \frac{u^{n+1}(1-n)}{n+1} = \boxed{\frac{(1+x^{1-n})^{n+1}(1-n)}{n+1} + C}$$

5.16 An Improper Loophole Log Integral

$$\int_0^{\infty} \sqrt[x]{x+1}\left(\frac{\ln(x+1)}{x^2} - \frac{1}{x(x+1)}\right) dx$$

If you experiment by deriving $(x+1)^{\frac{1}{x}}$, you'll soon realize that the antiderivative is $(x+1)^{\frac{1}{x}}$ itself. So now the tricky part is doing the improper bounds. After doing the limits carefully, your answer should be $\boxed{e-1}$.

5.17 A Crazy Infinite Logarithm Stair

$$\int_0^1 \frac{\ln(x) + \frac{\ln^2(x)+\frac{\ln^3(x)+\dots}{4}}{3}}{2} dx$$

$$= \int_0^1 \frac{\ln(x)}{2} + \frac{\ln^2(x)}{3!} + \frac{\ln^3(x)}{4!} + \dots dx$$

$$= \int_0^1 \sum_{n=2}^{\infty} \frac{\ln^{n-1}(x)}{n!} dx = \int_0^1 \left(\frac{x-1}{\ln(x)} - 1\right) dx = \boxed{\ln(2) - 1}$$

If you remember the famous Feynmann technique integral, you should get the following answer.

5.18 A Beautiful Secret with Two Infinite Functions

$$\int_0^{\frac{1}{2}} \frac{x}{x + \frac{x}{x+\dots}} + x^2 + x^3 + x^4 + \dots\, dx$$

Observe that

$$y = \frac{x}{x + \frac{x}{x+\dots}} = \frac{x}{x+y} \Rightarrow y^2 + xy = x \Rightarrow x = \frac{y^2}{1-y}.$$

$$y = x^2 + x^3 + x^4 + \dots = \frac{x^2}{1-x}.$$

This means that both functions are inverses of each other.

By sum of inverses, the integral equals to $\boxed{\frac{1}{4}}$.

The same concept can be applied with this integral.

$$\int_0^{\frac{1}{2}} \frac{x}{\sqrt{x + \frac{x}{\sqrt{x+\dots}}}} + \sqrt{x^3 + x^2\sqrt{x^3 + x^2\sqrt{x^3 + \dots}}} dx$$

5.19 An Awkward Denominator with Exponentials

$$\int_{-\infty}^{\infty} \frac{dx}{e^{x^2+x} + e^{x^2}}$$

Divide top and bottom by e^{x^2}:

$$= \int_{-\infty}^{\infty} \frac{e^{-x^2}}{e^x + 1} dx$$

Let $u = -x$:

$$= \int_{-\infty}^{\infty} \frac{e^{-u^2}}{e^{-u} + 1} du = \int_{-\infty}^{\infty} \frac{e^{-u^2} e^u}{e^u + 1} du$$

$$2I = \int_{-\infty}^{\infty} \frac{e^{-x^2}(e^x + 1)}{e^x + 1} dx = \sqrt{\pi}$$

So the answer becomes $\boxed{\frac{\sqrt{\pi}}{2}}$.

5.20 A Crazy Cube Root Integral

$$\int_0^7 \left(\sqrt[3]{\sqrt{x^2+1} + x} - \sqrt[3]{\sqrt{x^2+1} - x} \right) dx$$

Let $a = \sqrt[3]{\sqrt{x^2+1} + x}$ and $b = \sqrt[3]{\sqrt{x^2+1} - x}$. Let $y = a - b$ and compute y^3. Observe that

$$\begin{aligned} y^3 &= (a-b)^3 \\ &= a^3 - 3a^2b + 3ab^2 - b^3 \\ &= a^3 - b^3 - 3ab(a-b) \\ &= 2x - 3y. \end{aligned}$$

Notice that $x = \frac{y^3+3y}{2}$. We can instead integrate the inverse function since it is easier.

$$\int_a^b f(x)dx + \int_{f(a)}^{f(b)} f^{-1}(x)dx = bf(b) - af(a)$$

$$\int_0^7 ydx = 14 - \int_0^2 \frac{x^3 + 3x}{2} dx = \boxed{9}$$

5.21 An Evil Trig-Integral, Yet Easy Solution

$$\int_0^{\frac{\pi}{4}} \left(3\sec(x)\tan(x) + \sec(x)\tan^3(x)\right)^4 dx$$

$$\int_0^{\frac{\pi}{4}} \sec^4(x)\left(3\tan(x) + \tan^3(x)\right)^4 dx$$

Let $u = 3\tan(x) + \tan^3(x)$

$$= \frac{1}{3}\int_0^4 u^4 du = \boxed{\frac{1024}{15}}$$

5.22 What that 1/2 doin?

$$\int \left(\sinh(x)\tan^{-1}(e^x) + \frac{1}{2}\right) dx$$

Integrate by parts by deriving $\tan^{-1}(e^x)$ and integrating $\sinh(x)$:

$$= \tan^{-1}(e^x)\cosh(x) + \int \frac{1}{2} - \cosh(x) \cdot \frac{e^x}{1+e^{2x}} dx$$

$$= \tan^{-1}(e^x)\cosh(x) + \int \frac{1}{2} - \frac{e^x + e^{-x}}{2} \cdot \frac{dx}{e^x + e^{-x}} = \boxed{\tan^{-1}(e^x)\cosh(x) + C}$$

5.23 A Cool Infinite Exponential-Radical Function

$$\int e^{2x}\sqrt{e^{x^2}\sqrt[3]{e^{x^3}\sqrt[4]{e^{x^4}\sqrt[5]{e^{x^5}\ldots}}}}\, dx$$

$$= \int e^{2x} \cdot e^{\frac{x^2}{2} + \frac{x^3}{3!} + \frac{x^4}{4!} + \cdots} dx$$

Notice that by Taylor Series,

$$\sum_{n=1}^{\infty} \frac{x^n}{n!} = e^x - 1$$

Hence,

$$= \int e^x \cdot e^{e^x - 1} dx = \boxed{e^{e^x - 1} + C}$$

5.24 Multi-Logged Integral

$$\int_1^e \frac{\ln(x)\ln(\sqrt{2}x)\ln(2x)}{x}dx$$

Let $u = \ln(x)\ln(2x)$, then $du = \frac{\ln(2x)+\ln(x)}{x}dx = \frac{2\ln(\sqrt{2}x)}{x}dx$.

$$= \int_0^{1+\ln(2)} \frac{u}{2}du = \left[\frac{u^2}{4}\right]_0^{1+\ln(2)} = \boxed{\frac{(1+\ln(2))^2}{4}}$$

5.25 Exponential Beauty Distraction

$$\int \left(e^x - \frac{e^x}{x}\right)\left(e^{\frac{1}{x}} + \frac{e^{\frac{1}{x}}}{x}\right)dx$$

$$= \int e^x\left(1-\frac{1}{x}\right)e^{\frac{1}{x}}\left(1+\frac{1}{x}\right)dx = \int e^{x+\frac{1}{x}}\left(1-\frac{1}{x^2}\right)dx$$

Let $u = x + \frac{1}{x}$:

$$= \int e^u du = \boxed{e^{x+\frac{1}{x}} + C}$$

5.26 Decompose Under the Radical

$$\int \left(\frac{1}{(x+3)^2} - \frac{1}{(x+2)^2}\right)\frac{dx}{\sqrt{(x+2)(x+3)}}$$

Partial fractions under the radical:

$$= \int \left(\frac{1}{(x+3)^2} - \frac{1}{(x+2)^2}\right)\sqrt{\frac{1}{x+2} - \frac{1}{x+3}}dx$$

Let $u = \frac{1}{x+2} - \frac{1}{x+3}$, then

$$= \int \sqrt{u}du = \boxed{\frac{2}{3}\left(\frac{1}{(x+2)(x+3)}\right)^{\frac{3}{2}} + C}$$

5.27 Silver's Favorite Integral!!!

$$\int_1^{25} \lfloor \sqrt{x} \rfloor \bmod \lfloor x \rfloor \, dx$$

The easiest way to approach this is to list casework.
At $(1, 2)$, $1 \pmod 1 \equiv 0$.
At $(2, 4)$, $1 \pmod{2, 3, 4} \equiv 1$.
At $(4, 9)$, $2 \pmod{4, 5, ..., 9} \equiv 2$.
At $(9, 16)$, $3 \pmod{9, 10, ..., 16} \equiv 3$.
At $(16, 25)$, $4 \pmod{16, 17, ..., 25} \equiv 4$.

$$= \int_1^2 0 + \int_2^4 1 + \int_4^9 2 + \int_9^{16} 3 + \int_{16}^{25} 4 = \boxed{69}$$

5.28 Both Functions are Beautiful Together

$$\int_0^\infty e^{-x^{\frac{1}{3}}} x^{-\frac{1}{3}} + e^{-x^{\frac{2}{3}}} x^{-\frac{2}{3}} dx$$

For each integral:

$$\int_0^\infty e^{-x^{\frac{1}{3}}} x^{-\frac{1}{3}} dx \xrightarrow[3u^2 du = dx]{u = x^{\frac{1}{3}}} \int_0^\infty 3ue^{-u} du = 3$$

$$\int_0^\infty e^{-x^{\frac{2}{3}}} x^{-\frac{2}{3}} dx \xrightarrow{u = x^{\frac{1}{3}}} \int_0^\infty 3e^{-u^2} du = \frac{3\sqrt{\pi}}{2}$$

So our answer is $\boxed{3 + \frac{3\sqrt{\pi}}{2}}$.

5.29 A Second Derivative???

$$\int_1^e \left(\frac{d^2}{dx^2} [x^x] \right) \frac{dx}{x^x(1 + \ln(x))}$$

Notice that $x^x(1 + \ln(x)) = \frac{d}{dx}[x^x]$. Then let $u = x^x(1 + \ln(x))$:

$$= \int \frac{\frac{d^2}{dx^2}[x^x]}{\frac{d}{dx}[x^x]} dx = \int \frac{du}{u} = \ln |x^x(1 + \ln(x))|_1^e = \boxed{e + \ln(2)}$$

5.30 A Triggy Form of a Quartercircle Function

$$\int_0^{\sqrt{2}} \sec(\tan^{-1}(\cos(\sin^{-1}(x))))dx$$

Note that $\cos(\sin^{-1}(x)) = \sqrt{1-x^2}$ and $\sec(\tan^{-1}(x)) = \sqrt{x^2+1}$.

$$= \int_0^{\sqrt{2}} \sec(\tan^{-1}(\sqrt{1-x^2}))dx = \int_0^{\sqrt{2}} \sqrt{2-x^2}dx = \boxed{\frac{\pi}{2}}$$

5.31 Not Your Ordinary Trig-Sub Integral

$$\int_1^{\sqrt{2}} \frac{dx}{1-\sqrt{1-x^{-2}}}$$

Multiply top and bottom by x:

$$= \int_1^{\sqrt{2}} \frac{x}{x-\sqrt{x^2-1}}dx = \int_1^{\sqrt{2}} x(x+\sqrt{x^2-1})dx = \boxed{\frac{2\sqrt{2}}{3}}$$

5.32 A Delicious Way to have Pie

$$\int_0^1 \frac{1}{x}\sqrt{\frac{x^\pi}{1-x^\pi}}dx$$

Let $u = x^{\frac{\pi}{2}}$:

$$= \int_0^1 \frac{x^{\frac{\pi}{2}-1}}{\sqrt{1-x^\pi}}dx = \frac{2}{\pi}\int_0^1 \frac{du}{\sqrt{1-u^2}} = \boxed{1}$$

5.33 Ghost Symmetry with Arccosine!

$$\int_{-1}^1 \frac{\cos^{-1}(x)}{x^2+1}dx$$

Note that $\cos^{-1}(x) = \frac{\pi}{2} - \sin^{-1}(x)$. Then,

$$= \int_{-1}^1 \frac{\frac{\pi}{2}-\sin^{-1}(x)}{x^2+1}dx = \int_0^1 \frac{\pi}{1+x^2}dx = \boxed{\frac{\pi^2}{4}}$$

The $\sin^{-1}(x)$ is an odd function, so that terms goes to zero.

5.34 Not So Easy Now Is It...

$$\int_0^{\frac{\pi}{2}} \csc^2 x \cot x - \frac{1}{x^3} dx$$

$$= \left[\frac{1}{2x^2} - \frac{\cot^2(x)}{2}\right]_0^{\frac{\pi}{2}} = \frac{2}{\pi^2} - \lim_{x\to 0}\left(\frac{\tan^2(x) - x^2}{2x^2\tan^2(x)}\right) = \boxed{\frac{2}{\pi^2} - \frac{1}{3}}$$

That limit is ruthless...

5.35 That $\sqrt{2}$ is Very Necessary

$$\int_{\frac{\pi}{4}}^{\frac{3\pi}{4}} \sqrt{\sin(x) + \cos(x) + \sqrt{2}}\, dx$$

$$= \sqrt[4]{2}\int_{\frac{\pi}{4}}^{\frac{3\pi}{4}} \sqrt{\cos\left(x - \frac{\pi}{4}\right) + 1}\, dx = \sqrt[4]{2}\int_0^{\frac{\pi}{2}} \sqrt{\cos(x) + 1}\, dx = \boxed{2\sqrt[4]{2}}$$

5.36 Hyperbolic Cosine Hid the Frullani!!

$$\int_0^\infty \frac{1}{x}\ln\left(\frac{e^x + 1}{2\cosh(x)}\right) dx$$

Inside the logarithm, divide top and bottom by e^x, then

$$= \int_0^\infty \frac{1}{x}\ln\left(\frac{e^x + 1}{e^x + e^{-x}}\right) dx = \int_0^\infty \frac{\ln(1 + e^{-x}) - \ln(1 + e^{-2x})}{x} dx = \boxed{\ln^2(2)}$$

5.37 A Sneaky Gaussian Integral

$$\int_0^\infty x e^{-\left(x^2 - \frac{1}{x^2}\right)^2} dx$$

Let $u = \frac{1}{x}$:

$$\int_0^\infty \frac{e^{-(x^2 - \frac{1}{x^2})^2}}{x^3} dx \Rightarrow 2I = \int_0^\infty e^{-(x^2 - \frac{1}{x^2})^2}\left(x + \frac{1}{x^3}\right) dx$$

Let $u = x^2 - \frac{1}{x^2}$:

$$2I = \int_{-\infty}^\infty \frac{1}{2} e^{-u^2} du \Rightarrow I = \boxed{\frac{\sqrt{\pi}}{4}}$$

5.38 Holy Frullani Cosines!!

$$\int_0^\infty \frac{\cos(2x) - 4\cos(x) + 3}{x^3} dx$$

$$= \int_0^\infty \frac{1}{x}\left(\frac{\cos(2x) - 1 - 4\cos(x) + 4}{x^2}\right) dx$$

$$= \int_0^\infty \frac{1}{x}\left(\frac{4 - 4\cos(x) - (1 - \cos(2x))}{x^2}\right) dx$$

$$= \int_0^\infty \frac{4}{x}\left(\frac{1 - \cos(x)}{x^2} - \frac{1 - \cos(2x)}{4x^2}\right) dx$$

Use Frullani Integral:

$$4\ln\left(\frac{1}{2}\right)\left[\frac{1 - \cos(x)}{x^2}\right]_0^\infty = \boxed{\ln(4)}$$

5.39 A Sinh Trapped Between the Logs

$$\int_{-1}^1 \ln(\sinh^2(\ln(x)))dx$$

Note that $\sinh(\ln(x)) = \frac{1}{2}(x - \frac{1}{x})$.

$$= \int_{-1}^1 \ln\left(\left(x - \frac{1}{x}\right)^2\right) - \ln(4)dx$$

By symmetry,

$$= 4\int_0^1 \ln\left(\frac{1}{x} - x\right) dx - 4\ln(2)$$

Observe that,

$$\Rightarrow \int_0^1 \ln\left(\frac{1}{x} - x\right) dx = \int_0^1 \ln(1 + x) + \ln(1 - x) - \ln(x)dx$$

By King's Rule, the last two terms cancels out each other when the last terms applies $u = 1 - x$.

$$= \int_0^1 \ln(x + 1)dx = 2\ln(2) - 1.$$

Hence,

$$4(2\ln(2) - 1) - 4\ln(2) = \boxed{4\ln(2) - 4}$$

5.40 An Odd Way to Write a Log Integral

$$\int (x+1)(x+\ln(x))\left(1+\frac{\ln(x)}{x}\right)dx$$

Factor x out from $(x+1)$ and move the x to $(1+\frac{\ln(x)}{x})$.

$$=\int\left(1+\frac{1}{x}\right)(x+\ln(x))^2dx$$

Let $u=x+\ln(x)$:

$$=\int u^2du=\frac{u^3}{3}=\boxed{\frac{1}{3}(x+\ln(x))^3+C}$$

5.41 Where's the Upper Infinity???

$$\int_{-\infty}^{0}\frac{2^x+\frac{1}{2}}{2^x+2^{-x}+1}dx$$

This is actually very tricky, multiply top and bottom by 2^x:

$$=\int_{-\infty}^{0}\frac{4^x+2^{x-1}}{4^x+2^x+1}dx$$

Let $u=4^x+2^x+1$, then

$$\begin{aligned}du &= 4^x\ln(4)+2^x\ln(2)\ dx\\ &= 4^x 2\ln(2)+2^x\ln(2)\ dx\\ &= \ln(2)(2\cdot 4^x+2^x)\ dx\\ &= 2\ln(2)(4^x+2^{x-1})\ dx.\end{aligned}$$

Thus, we have

$$=\frac{1}{\ln(4)}\int_1^3\frac{du}{u}=\boxed{\frac{\ln(3)}{\ln(4)}}$$

5.42 Consecutive Constants!!!

$$\int_0^\infty \frac{(2x+5)}{(x+1)(x+2)(x+3)(x+4)}dx$$

$$= \int_0^\infty \frac{(2x+5)}{[(x+1)(x+4)][(x+2)(x+3)]}dx = \int_0^\infty \frac{2x+5}{(x^2+5x+4)(x^2+5x+6)}dx$$

Let $u = x^2+5x+4$, then $du = (2x+5)dx$:

$$= \int_4^\infty \frac{du}{u(u+2)} = \frac{1}{2}\int_4^\infty \frac{1}{u} - \frac{1}{u+2}du = \frac{1}{2}\left[\ln|u| - \ln|u+2|\right]_4^\infty = \boxed{\frac{1}{2}\ln\left(\frac{3}{2}\right)}$$

Note:

$$\lim_{x\to\infty}[\ln(x) - \ln(x+2)] = \lim_{x\to\infty}\ln\left(\frac{x}{x+2}\right) = \ln\left(\lim_{x\to\infty}\frac{x}{x+2}\right) = \ln(1) = 0$$

5.43 A Chaotic Arctangent

$$\int_{-1}^1 x^2 \tan^{-1}\left(\frac{e^x - x}{1+xe^x}\right)dx$$

This is a very sneaky ninja substitution. Let $u = -x$.
Again, we can just change the variables back.

$$= \int_{-1}^1 x^2\tan^{-1}\left(\frac{e^{-x}+x}{1-xe^{-x}}\right)dx = \int_{-1}^1 x^2\tan^{-1}\left(\frac{1+xe^x}{e^x-x}\right)dx$$

We now have

$$I = \int_{-1}^1 x^2\tan^{-1}\left(\frac{e^x-x}{1+xe^x}\right)dx = \int_{-1}^1 x^2\tan^{-1}\left(\frac{1+xe^x}{e^x-x}\right)dx$$

There is a well-known identity where for $f(x) > 0$,

$$\tan^{-1}(f(x)) + \tan^{-1}\left(\frac{1}{f(x)}\right) = \frac{\pi}{2}.$$

Luckily for all $x \in (-1,1)$, $\frac{e^x-x}{1+xe^x} > 0$. So when we add both of the same integrals in two different forms, we get

$$2I = \int_{-1}^1 x^2 \cdot \frac{\pi}{2}dx = \frac{\pi}{3}.$$

Therefore, our answer is

$$I = \boxed{\frac{\pi}{6}}$$

5.44 Composite Inverse Functions???

$$\int_0^{\frac{\pi}{2}} \frac{\pi}{2}\sin\left(\frac{\pi}{2}(1-\sin(x))\right)+\cos^{-1}\left(\frac{2}{\pi}\cos^{-1}\left(\frac{2}{\pi}x\right)\right)dx$$

Let $f(x)=\frac{\pi}{2}\sin(\frac{\pi}{2}(1-\sin(x)))$. Then $f^{-1}(x)=\sin^{-1}(1-\frac{2}{\pi}\sin^{-1}(\frac{2}{\pi}x))$. However, note that $\cos^{-1}(x)=\frac{\pi}{2}-\sin^{-1}(x)$, so

$$\begin{aligned}&\cos^{-1}\left(\frac{2}{\pi}\cos^{-1}\left(\frac{2}{\pi}x\right)\right)\\&=\frac{\pi}{2}-\sin^{-1}\left(\frac{2}{\pi}\cos^{-1}\left(\frac{2}{\pi}x\right)\right)\\&=\frac{\pi}{2}-\sin^{-1}\left(1-\frac{2}{\pi}\sin^{-1}\left(\frac{2}{\pi}x\right)\right).\end{aligned}$$

Hence,

$$=\int_0^{\frac{\pi}{2}} \frac{\pi}{2}\sin\left(\frac{\pi}{2}(1-\sin(x))\right)+\frac{\pi}{2}-\sin^{-1}\left(1-\frac{2}{\pi}\sin^{-1}\left(\frac{2}{\pi}x\right)\right)dx$$

There is a famous formula where for any function $f(x)$,

$$\int_a^b f(x)dx+\int_{f(a)}^{f(b)} f^{-1}(x)dx=bf(b)-af(a).$$

Observe that for $f(x)=\frac{\pi}{2}\sin(\frac{\pi}{2}(1-\sin(x)))$, $f(0)=\frac{\pi}{2}$ and $f(\frac{\pi}{2})=0$, so

$$\int_0^{\frac{\pi}{2}} f(x)dx+\int_{\frac{\pi}{2}}^0 f^{-1}(x)dx=0=\int_0^{\frac{\pi}{2}} f(x)dx-\int_0^{\frac{\pi}{2}} f^{-1}(x)dx$$

Therefore,

$$=\int_0^{\frac{\pi}{2}} \frac{\pi}{2}\sin\left(\frac{\pi}{2}(1-\sin(x))\right)+\frac{\pi}{2}-\sin^{-1}\left(1-\frac{2}{\pi}\sin^{-1}\left(\frac{2}{\pi}x\right)\right)dx=\int_0^{\frac{\pi}{2}}\frac{\pi}{2}dx=\boxed{\frac{\pi^2}{4}}$$

5.45 Something's Not Right Here...

$$\int \ln\left(3x^5+5x^3\right)\left(\frac{x^2+1}{3x^3+5x}\right)dx$$

Multiply x^2 top and bottom, then $u=3x^5+5x^3$:

$$=\int \ln\left(3x^5+5x^3\right)\left(\frac{x^4+x^2}{3x^5+5x^3}\right)dx=\frac{1}{15}\int\frac{\ln(u)}{u}du$$

$$=\boxed{\frac{1}{30}\ln^2(3x^5+5x^3)+C}$$

5.46 A Crazy Infinite Log Function

$$\int_0^2 x\sqrt{x^{\ln(x)}\sqrt[3]{x^{\ln^2(x)}\sqrt[4]{x^{\ln^3(x)}\sqrt[5]{\cdots}}}}dx$$

$$= \int_0^2 x^{1+\frac{\log(x)}{2}+\frac{\log^2(x)}{3!}+\frac{\log^3(x)}{4!}+\cdots}dx = \int_0^2 x^{\sum_{n=1}^{\infty}\frac{\log^{n-1}(x)}{n!}}dx$$

By Taylor Series,

$$\sum_{n=1}^{\infty}\frac{\log^{n-1}(x)}{n!} = \frac{1}{\log(x)}\sum_{n=1}^{\infty}\frac{\log^n(x)}{n!} = \frac{x-1}{\log(x)}.$$

$$= \int_0^2 x^{\frac{x-1}{\log(x)}}dx = \int_0^2 e^{x-1}dx = \boxed{e-\frac{1}{e}}$$

5.47 A Disgusting IBP Trap

$$\int_0^{e^{2\pi}} \cos(\log(x))\log(x)dx$$

Complexifying the integral might be the fastest.

$$\Re\left(\int_0^{e^{2\pi}} x^i\log(x)dx\right) = \Re\left(\left[\frac{ix^{i+1}}{2}+\left(\frac{1}{2}-\frac{i}{2}\right)x^{1+i}\log(x)\right]_0^{e^{2\pi}}\right)$$

$$= \Re\left(\left[\frac{ie^{2\pi i+2\pi}}{2}-0+\left(\frac{1}{2}-\frac{i}{2}\right)e^{2\pi+2\pi i}(2\pi)-0\right]\right) = \left(\frac{1}{2}e^{2\pi+2\pi i}(2\pi)\right) = \boxed{\pi e^{2\pi}}$$

5.48 A Gaussian Imposter

$$\int_{-1}^{\infty} e^{-x^2}(5x^4-2x^6)dx$$

Very hard to see, but it's actually reverse product rule.

$$\left[x^5e^{-x^2}\right]_{-1}^{\infty} = \boxed{\frac{1}{e}}$$

5.49 A Hyperbolic Ninja

$$\int_{-\frac{\pi}{3}}^{\frac{\pi}{3}} \frac{e^x \sec^2(x)}{\cosh(x)} dx$$

Let $u = -x$:

$$= \frac{1}{2}\int_{-\frac{\pi}{3}}^{\frac{\pi}{3}} \frac{(e^x + e^{-x})\sec^2(x)}{\cosh(x)} dx = 2\int_0^{\frac{\pi}{3}} \sec^2(x)dx = [2\tan(x)]_0^{\frac{\pi}{3}} = \boxed{2\sqrt{3}}$$

5.50 An Obscure King!

$$\int_0^1 \frac{e^x\sqrt{1-x}}{e^x\sqrt{1-x}+\sqrt{ex}} dx$$

It's just secretly King's Rule, let $u = 1 - x$.

$$= \int_0^1 \frac{e^{1-x}\sqrt{x}}{e^{1-x}\sqrt{x}+\sqrt{e(1-x)}} dx = \int_0^1 \frac{\sqrt{ex}}{\sqrt{ex}+e^x\sqrt{1-x}} dx \Rightarrow \frac{1}{2}\int_0^1 1dx = \boxed{\frac{1}{2}}$$

5.51 An Alternating Infinite Trig Function

$$\int_0^{\frac{\pi}{2}} \sqrt{\sin(x)\sqrt{\tan(x)\sqrt{\sin(x)\sqrt{\tan(x)\sqrt{\dots}}}}}\, dx$$

Observe that

$$\begin{aligned}
w &= \sqrt{\sin(x)\sqrt{\tan(x)w}} \\
w^2 &= \sin(x)\sqrt{\tan(x)w} \\
w^4 &= \sin^2(x)\tan(x)w \\
w^3 &= \sin^2(x)\tan(x) = \frac{\sin^3(x)}{\cos(x)} \\
w &= \frac{\sin(x)}{\sqrt[3]{\cos(x)}}.
\end{aligned}$$

$$= \int_0^{\frac{\pi}{2}} \frac{\sin(x)}{\sqrt[3]{\cos(x)}} dx = \left[-\frac{3}{2}\cos^{2/3}(x)\right]_0^{\frac{\pi}{2}} = \boxed{\frac{3}{2}}$$

5.52 A Scary Radical Tangent Integral

$$\int_0^{\frac{\pi}{2}} \sqrt{\tan(x)+\cot(x)+2}\, dx$$

$$= \int_0^{\frac{\pi}{2}} \sqrt{2+\frac{1}{\sin(x)\cos(x)}}dx = \int_0^{\frac{\pi}{2}} \frac{\sqrt{2}(\sin(x)+\cos(x))}{\sqrt{\sin(2x)}}dx = \int_0^{\frac{\pi}{2}} \frac{\sqrt{2}(\sin(x)+\cos(x))}{\sqrt{1-(\sin(x)-\cos(x))^2}}dx$$

BE VERY CAREFUL WITH THE BOUNDS AND ABSOLUTE VALUES FROM SQUARE ROOTS!!!

$$= \int_0^{\frac{\pi}{4}} \ldots + \int_{\frac{\pi}{4}}^{\frac{\pi}{2}} \ldots dx = 2\int_0^1 \frac{\sqrt{2}}{\sqrt{1-u^2}}du = \boxed{\sqrt{2}\pi}$$

5.53 What we doin with that radical???

$$\int \sqrt{\frac{e^x}{e^x-x}}\left(1-\frac{1}{x}\right) dx$$

$$= \int \frac{1}{\sqrt{1-xe^{-x}}} \cdot \frac{(x-1)e^{-x}}{xe^{-x}}dx$$

Let $u = xe^{-x}$, $du = (e^{-x} - xe^{-x})dx = (1-x)e^{-x}dx$:

$$= \int \frac{-du}{u\sqrt{1-u}} = \int \frac{2wdw}{(1-w^2)w} = 2\tanh^{-1}(w) = \boxed{2\tanh^{-1}(\sqrt{1-xe^{-x}})+C}$$

5.54 Slickest Logarithmic Manuever

$$\int_1^e \ln\left(x^{x^x}\right)(x+\ln(x^x))dx$$

$$= \int_1^e x^x\ln(x)\cdot x(1+\ln(x))dx = \int_1^e \ln(x^x)x^x(1+\ln(x))dx$$

Let $u = x^x$:

$$= \int_1^{e^e} \ln(u)du = [u\ln(u)-u]_1^{e^e} = \boxed{e^e(e-1)+1}$$

5.55 A Shocking Simplification

$$\int_1^\infty \frac{\sqrt[x]{2}}{x^2 + \sqrt[x]{2x^{2x}}} dx$$

As horrendous as this looks, this is actually fairly a simple integral. All it takes is just a tiny algebraic manipulation.

$$= \int_1^\infty \frac{2^{1/x}}{x^2 + 2^{1/x} \cdot x^2} dx = \int_1^\infty \frac{2^{1/x} dx}{x^2(1 + 2^{1/x})}$$

Let $u = \frac{1}{x}$:

$$= \int_0^1 \frac{2^u}{2^u + 1} du = \boxed{\log_2(3) - 1}$$

5.56 A Super Sneaky Trig-Sub

$$\int_{-1}^1 \frac{dx}{\sqrt{1+x} + \sqrt{1-x} + 2}$$

Let $x = \sin(2\theta)$. This is because $1 \pm \sin(2\theta) = (\sin\theta \pm \cos\theta)^2$.

$$= \int_{-\frac{\pi}{4}}^{\frac{\pi}{4}} \frac{2\cos(2\theta) d\theta}{\sin\theta + \cos\theta + |\sin\theta - \cos\theta| + 2} = 2\int_0^{\frac{\pi}{4}} \frac{\cos(2\theta)}{\cos\theta + 1} d\theta$$

$$= 2\int_0^{\frac{\pi}{4}} \frac{1 - 2\sin^2\theta}{1 + \cos\theta} d\theta = \int_0^{\frac{\pi}{4}} \frac{2}{1 + \cos\theta} - 4(1 - \cos\theta) d\theta$$

$$= \int_0^{\frac{\pi}{4}} (2\csc^2\theta - 2\csc\theta\cot\theta - 4 + 4\cos\theta) d\theta = [2\csc\theta - 2\cot\theta - 4\theta + 4\sin\theta]_0^{\frac{\pi}{4}}$$

$$= 2\sqrt{2} - 2 - \pi + \frac{4}{\sqrt{2}} - \lim_{\theta \to 0} \frac{1 - \cos\theta}{\sin\theta} = \boxed{4\sqrt{2} - 2 - \pi}$$

Chapter 6

Solutions for Problems made for Training

6.1 The Golden Trig Integral!!!

$$\int_{\frac{\pi}{3}}^{\frac{\pi}{2}} \frac{\tan(x)}{\sqrt{2+\tan^2(x)}}dx$$

$$= \int_{\frac{\pi}{3}}^{\frac{\pi}{2}} \frac{\sec(x)\tan(x)}{\sec(x)\sqrt{1+\sec^2(x)}}dx = \int_2^{\infty} \frac{du}{u\sqrt{1+u^2}} = \int_{\tan^{-1}(2)}^{\frac{\pi}{2}} \csc\theta d\theta$$

$$= -\ln|\csc\theta + \cot\theta|_{\tan^{-1}(2)}^{\frac{\pi}{2}} = \ln\left(\frac{1}{2} + \csc(\tan^{-1}(2))\right) = \boxed{\ln(\varphi)}$$

6.2 Lost in the Polynomials

$$\int (x^2+x)^2(x^2-x+1)^2dx$$

$$= \int x^2(x+1)^2(x^2-x+1)^2dx = \int x^2(x^3+1)^2dx = \boxed{\frac{1}{9}(x^3+1)^3+C}$$

6.3 An Awkward Formatted Integral

$$\int \frac{dx}{(x+1)\sqrt{x}+2x}$$

$$= \int \frac{dx}{\sqrt{x}(x+2\sqrt{x}+1)} \xrightarrow{u=1+\sqrt{x}} \int \frac{2}{u^2}du = \boxed{-\frac{2}{1+\sqrt{x}}+C}$$

6.4 Just a Minor Sign Flip

$$\int_{\frac{1}{4}}^{\frac{1}{2}} \frac{\sqrt{x-\sqrt{x-\frac{1}{4}}}}{\sqrt{x+\sqrt{x-\frac{1}{4}}}}dx$$

Let $u = \sqrt{x-\frac{1}{4}}$:

$$= \int_0^{1/2} \frac{\sqrt{u^2+\frac{1}{4}-u}}{\sqrt{u^2+\frac{1}{4}+u}} \cdot 2udu = \int_0^{1/2} \frac{|u-\frac{1}{2}|}{u+\frac{1}{2}}2udu = \int_0^{1/2} \frac{1-2u}{2u+1} \cdot 2udu$$

$$= \int_0^{1/2} 2-2u-\frac{2}{2u+1}du = \boxed{\frac{3}{4}-\ln(2)}$$

6.5 A Very Misleading Simple Integral

$$\int_1^{\sqrt{2}} (x^2+1)^3\left(x+\frac{1}{x^9}\right)dx$$

If you've never seen this integral before, you can be stuck and confuzzled on how to speed integrate this. However, the simple solution may frustrate you. All you had to do was to separate into two terms:

$$= \int_1^{\sqrt{2}} x(x^2+1)^3 + \frac{(x^2+1)^3}{x^9}dx = \int_2^3 \frac{1}{2}u^3du + \int_1^{\sqrt{2}} \left(1+\frac{1}{x^2}\right)^3 \frac{dx}{x^3}$$

$$= \left[\frac{1}{8}u^4\right]_2^3 + \int_{3/2}^2 \frac{u^3}{2}du = \left[\frac{u^4}{8}\right]_{3/2}^3 = \boxed{\frac{81}{8}-\frac{81}{128}}$$

6.6 A Double Reverse Product Rule Hidden Inside

$$\int \cos(x)(x^3+6x)dx$$

You can choose the normal way, which is integration by parts. OORRRRR, you can choose the quicker way which is letting $0=\sin(x)(3x^2-3x^2)$.

$$=\int x^3\cos(x)+3x^2\sin(x)+6x\cos(x)-3x^2\sin(x)dx=\boxed{x^3\sin(x)+3x^2\cos(x)+C}$$

6.7 A Chaotic Complexifying Speed Bash

$$\int e^{e^x+x}\left(\cos\left(e^{e^x}\right)+\cos(e^x)\right)dx$$

$$=\Re\left(\int e^{e^x+x}\left(e^{ie^{e^x}}+e^{ie^x}\right)dx\right)\xrightarrow[t=e^x]{u=e^{e^x}}\Re\left(\int e^{iu}du+\int e^{t(1+i)}dt\right)$$

$$=\Re\left(\frac{e^{iu}}{i}+\frac{e^{t(1+i)}}{1+i}=-i(\cos(u)+i\sin(u))+\frac{(1-i)}{2}e^t(\cos(t)+i\sin(t))\right)$$

$$=\sin(u)+\frac{e^t}{2}\cos(t)+\frac{e^t}{2}\sin(t)=\boxed{\sin\left(e^{e^x}\right)+\frac{e^{e^x}}{2}(\cos(e^x)+\sin(e^x))+C}$$

6.8 Don't You Dare Use Feynman...

$$\int_0^{\ln(2)}\frac{\sec^{-1}(\cosh(x))}{\cosh(x)}dx$$

Literally, you do not need Feynman technique to solve this. And even if you want to, I can punish you by removing the bounds (yes, this integral is elementary). Let $u=\cosh(x)$:

$$=\int_1^{5/4}\frac{\sec^{-1}(u)}{u\sqrt{u^2-1}}du=\left[\frac{1}{2}(\sec^{-1}(u))^2\right]_1^{5/4}=\boxed{\frac{1}{2}\left(\sec^{-1}\left(\frac{5}{4}\right)\right)^2}$$

6.9 The 8th Power Won't Save You...

$$\int_0^{\infty} \frac{dx}{(1+x)(1+x^2)(1+x^4)}$$

Let $u = \frac{1}{x}$:

$$= \frac{1}{2}\int_0^{\infty} \frac{x^5+1}{(x+1)(x^2+1)(x^4+1)}dx = \frac{1}{2}\int_0^{\infty} \frac{x^4-x^3+x^2-x+1}{(x^2+1)(x^4+1)}dx$$

$$= \frac{1}{2}\int_0^{\infty} \frac{(x^4+1)-x(x^2+1)+x^2}{(x^2+1)(x^4+1)}dx = \frac{1}{2}\int_0^{\infty} \frac{1}{x^2+1} - \frac{x}{x^4+1} + \frac{1}{x^4+1} - \frac{1}{(x^2+1)(x^4+1)}dx$$

The first two terms are easy peasy.
The third term can be solved with beta function or Glasser's Master Theorem.
The fourth term can be solved with $x = \tan\theta$ and then Queen's rule.

$$= \frac{1}{2}\left[\frac{\pi}{2} - \frac{1}{2}\left(\frac{\pi}{2}\right) + \frac{\pi}{2\sqrt{2}} - \frac{\pi}{4}\right] = \boxed{\frac{\pi}{4\sqrt{2}}}$$

6.10 Absorb the Exponential Function

$$\int \sinh^5(x)e^{7x}dx$$

$$= \int \left(\frac{e^x - e^{-x}}{2}\right)^5 e^{7x}dx = \frac{1}{2^5}\int (e^{2x}-1)^5 e^{2x}dx = \boxed{\frac{(e^{2x}-1)^6}{6\cdot 2^6} + C}$$

6.11 King Tangent!!!

$$\int_0^1 \frac{\tan(x)}{1-\tan(x)\tan(1-x)}dx$$

Let $u = 1 - x$, then

$$= \frac{1}{2}\int_0^1 \frac{\tan(x)+\tan(1-x)}{1-\tan(x)\tan(1-x)}dx = \frac{1}{2}\int_0^1 \tan(1)dx = \boxed{\frac{\tan(1)}{2}}$$

6.12 A Quirky Loophole Log Integral

$$\int_0^\infty \frac{\ln(x)(\ln(x)+1)}{x^{2x-1}}dx$$

$$= \int_0^\infty \frac{\ln(x)(\ln(x)+1)}{e^{(2x-1)\ln(x)}}dx = \int_0^\infty \frac{x\ln(x)(\ln(x)+1)}{e^{2x\ln(x)}}dx$$

Let $u = x\ln(x)$:

$$= \int_0^\infty ue^{-2u}du = \boxed{\frac{1}{4}}$$

6.13 Monomial Substitution

$$\int_{-4}^{-3} \frac{(x+4)^3}{(x+3)^2(x+5)^2+1}dx$$

$$= \int_{-4}^{-3} \frac{(x+4)^3}{(x^2+8x+15)^2+1}dx = \int_0^1 \frac{u^3}{(u^2-1)^2+1}du = \frac{1}{2}\int_0^1 \frac{w+1}{w^2+1}dw = \boxed{\frac{\pi}{8}+\frac{\ln(2)}{4}}$$

6.14 Borwein???

$$\int_0^\infty \sin\left(\frac{1}{x}\right)\sin\left(\frac{1}{3x}\right)dx$$

$$= \int_0^\infty \frac{\sin(x)\sin(\frac{x}{3})}{x^2}dx = \int_0^\infty \frac{\sin(x)}{x}\cdot\frac{\sin(\frac{x}{3})}{\frac{x}{3}}\cdot\frac{dx}{3} = \boxed{\frac{\pi}{6}}$$

By Borwein Integrals:

$$\int_0^\infty \frac{\sin(x)}{x}\cdot\frac{\sin(x/3)}{x/3}dx = \int_0^\infty \frac{\sin(x)}{x}\cdot\frac{\sin(x/3)}{x/3}\cdot\frac{\sin(x/5)}{x/5}dx = \ldots = \int_0^\infty \prod_{k=1}^{7}\frac{\sin(x/(2k-1))}{x/(2k-1)}dx = \frac{\pi}{2}$$

6.15 Anti-Conjugate!!!

$$\int_0^1 \frac{x-\sqrt{x-x^2}}{2x-1}dx$$

$$= \int_0^1 \frac{x^2-(x-x^2)}{(2x-1)(x+\sqrt{x-x^2})}dx = \int_0^1 \frac{x}{x+\sqrt{x-x^2}}dx = \int_0^1 \frac{\sqrt{x}}{\sqrt{x}+\sqrt{1-x}}dx = \boxed{\frac{1}{2}}$$

6.16 A Deadly Polynomial-Log Integral

$$\int x(\ln^2(x)+1)(\ln(x)+1)^2dx$$

The sneakiest thing about polynomial logarithms is that deriving $xP(\ln(x))$ gives you a pure polynomial in terms of $\ln(x)$. In this case, let $u = x(\ln^2(x)+1)$, then $du = (1+\ln(x))^2dx$.

$$= \int udu = \boxed{\frac{1}{2}(x\ln^2|x|+x)^2+C}$$

6.17 Ugliest Chaotic Integral Monster

$$\int_0^2 \frac{3x^3}{2\sqrt{1+x^3}} + \frac{2x^2+2x}{3(x^2+2x)^{\frac{2}{3}}}dx$$

Both terms are quite suspicious, let's do IBP on each term.
For first term, integrate $\frac{3x^2}{2\sqrt{1+x^3}}$ and for second term, integrate $\frac{1}{3}(2x+2)(x^2+2x)^{-2/3}$.

$$= \left[x\sqrt{x^3+1}+x\sqrt[3]{x^2+2x}\right]_0^2 - \int_0^2 \sqrt{x^3+1}+\sqrt[3]{x^2+2x}\,dx$$

$$= 10 - \int_0^2 \sqrt{x^3+1}+\sqrt[3]{(x+1)^2-1}\,dx = 10 - \int_0^2 \sqrt{x^3+1}dx - \int_1^3 \sqrt[3]{x^2-1}dx$$

The last two integrals are sum of inverses. Thus, the answer should be $10-2(3) = \boxed{4}$.

6.18 Anti-Conjugate Again!!!

$$\int_0^1 \frac{2x-1}{\sqrt{2x-x^2}-\sqrt{1-x^2}}dx$$

$$= \int_0^1 \frac{(2x-1)(\sqrt{2x-x^2}+\sqrt{1-x^2})}{2x-x^2-1+x^2}dx = \int_0^1 \sqrt{2x-x^2}+\sqrt{1-x^2}dx = \boxed{\frac{\pi}{2}}$$

The last integral is simply area of partial circles.

6.19 Time for Taylor Series!!

$$\int_0^\infty xe^{-x}\operatorname{sech}(x)dx$$

$$= \int_0^\infty \frac{2x}{e^{2x}+1}dx = \frac{1}{2}\int_0^\infty \frac{ue^{-u}}{e^{-u}+1}du = \frac{1}{2}\sum_{n=1}^\infty (-1)^n \int_0^\infty -ue^{-un}du$$

$$= \frac{1}{2}\sum_{n=1}^\infty (-1)^n\left(-\frac{1}{n^2}\right) = -\frac{1}{2}\sum_{n=1}^\infty \frac{(-1)^n}{n^2} = \boxed{\frac{\pi^2}{24}}$$

What About Me?

Wasabi!! (Wassup) I am Yuepeng Alex Yang, mostly known as Silver on the internet. I've been speed integrating since 11th grade and started writing my own integration bee problems in 12th grade. During the time, integration bees were not very popular, and I wished that it was. Since there weren't that many competitions that share their integrals, it was hard to find newer integration bee problems. In addition, I scold any universities who don't share their past integration bee problems because I believe they just reuse them. That's where I decided to make my own integrals and recorded very cool discoveries and unimaginable integration techniques.

I've actually gotten into competitive math since 7th grade and joined Mathorama. However in my 8th grade year, I completely stopped due to bullying and the toxic elitism in my school. I didn't feel comfortable around any "mathletes", so I quit pursuing in math and began majoring in Chemistry. Because of this, I've never competed in an actual math contest in my life. I remember during high school, my math teacher would always begged me to compete at a Math Circle, but then again, I just wasn't comfortable being in a group of mathletes.

During 12th grade, I debated to compete at an integration bee competition. Because it was mostly undergrads competing, I decided to give it a try. Thankfully, it wasn't too bad and I was okay being there since my friends were with me. Not only that, there were rarely any high school students competing with me; the majority were undergrads. Unfortunately in my first integration bee competition, I got eliminated at semi-finals. I was going too fast which made me become inaccurate with my answers; a complete lesson learned. Sadly, integration bees are annual, so at the end of my 12th grade year, I kept coming up with more cool integration bee problems as well as trained an amazing friend of mine who later became top 16 qualifier for MIT Integration Bee 2022. (So proud of her!!!) During my freshmen undergrad, I started problem writing for many integration bees and math tournaments, as well as winning three of them in my hometown: first from Reedley College and the rest from my own school, CSU Fresno, 2-years in a row both perfectly scored in all rounds. I wanted to win more, but the pandemic hit sadly. Ironically, I switched late being a math major, and thankfully finished in only two years. Now, I pursue in becoming a math professor for a community college or a university, and currently focus on popularizing speed integration. If no one wants to share all the dark forbidden arts of integration techniques, then I will.

www.ingramcontent.com/pod-product-compliance
Lightning Source LLC
LaVergne TN
LVHW082359200926
848481LV00012B/3622
* 9 7 9 8 2 3 1 9 4 8 8 0 2 *